THE WORLD

IN MINIATURE;

EDITED BY

FREDERIC SHOBERL

SOUTH SEA ISLANDS

BEING

A DESCRIPTION

OF THE

MANNERS, CUSTOMS, CHARACTER, RELIGION,

AND STATE OF SOCIETY

AMONG THE VARIOUS TRIBES SCATTERED OVER

THE GREAT OCEAN, CALLED THE PACIFIC,

OR THE SOUTH SEA

VOLUME I

Elibron Classics

www.elibron.com

Elibron Classics series.

ISBN 1-4212-2575-1 (paperback)
ISBN 1-4212-2574-3 (hardcover)

This Elibron Classics Replica Edition is an unabridged facsimile of the edition published in 1820 by Frederic Shoberl, London.

Frontispiece to Vol. I.

Dance of the Friendly Islands.

THE WORLD

IN MINIATURE;

EDITED BY

FREDERIC SHOBERL.

South Sea Islands;

BEING

A DESCRIPTION

OF THE

MANNERS, CUSTOMS, CHARACTER, RELIGION,

AND STATE OF SOCIETY

AMONG THE VARIOUS TRIBES SCATTERED OVER

The GREAT OCEAN, *called the* PACIFIC,

or the SOUTH SEA:

ILLUSTRATED WITH

Twenty-Six Coloured Engravings

IN TWO VOLUMES.

VOL. I.

"The proper study of mankind is man."---POPE.

LONDON:

PRINTED FOR R. ACKERMANN, REPOSITORY

OF ARTS, STRAND;

And to be had of all Booksellers.

LONDON:
Green, Leicester Street, Leicester Square.

PREFACE.

There is not perhaps any portion of the globe that presents, at this moment, a spectacle so full of interest to the contemplative mind as the islands scattered over the vast ocean which intervenes between the Asiatic and American continents. Half a century ago, many of these islands were scarcely known even by name to the civilized world; and most of them, though indeed casually seen by earlier mariners, had never been explored by Europeans, till the peaceful expeditions equipped by the British government in the early part of the reign of the late king, and the indefatigable researches of our great navigator, Captain Cook, exhibited their inhabitants in all the freshness of a new discovery, and opened to the philosopher a fertile theme of enquiry and speculation.

In all these tribes, how distant soever from each other, much the same habits and manners and a nearly equal degree of barbarism were found to prevail. With the benevolent design of improving their condition, the officers of our ships industriously introduced among them the most serviceable of our domestic animals to which they were utter strangers, and such generally useful vegetable productions as were best adapted to the soil and climate of their respective islands. The intercourse with Europeans, which from this time became more frequent, served to make these savages acquainted with the superiority of their visitors in all those arts that tend to the preservation and embellishment of life. An eager desire to possess themselves of our mechanical instruments, and a spirit of imitation were the natural consequences of this impression. The change thus gradually operating among them was accelerated by the establishment in some of the islands of missionaries, whose religious labours, however, seemed for a long series of years to be totally fruitless. Their perseverance has nevertheless been crowned with a result surpassing the most sanguine expectations ; and a revolution which, we trust, will extend over the whole of the Great Ocean, is now in rapid progress among some of its tribes. Among savages who, a short time since, were but a few degrees removed from a state of nature, printing presses have been established—written laws have been promulgated—the trial by jury adopted—the rudiments of navies formed—regular roads made—churches built—societies for the dissemination of the Scriptures and the encouragement of the Arts instituted—and the atrocious cruelties of the ancient superstition have yielded to the beneficent influence of the Gospel of Christ!

Sincerely as we should rejoice in such a change, by whomsoever effected, we must confess that it heightens our gratification to find such wonders accomplished through the instrumentality of Englishmen, and much of a British spirit and British feelings diffused along with these improvements. The extension of that spirit and of those feelings to the remotest corners of the globe we hail with cordial exultation, not merely because England is the land of our birth, but because we are convinced that institutions arising out of them are better calculated to promote the liberty, prosperity and happiness of mankind than those of any other nation under the face of heaven.

In justice to the authors of some

of the most recent publications re-specting the South Sea Islands, we cannot omit acknowledging the assistance derived from them in the compilation of these volumes. Our first tribute is due to the valuable work of Mr. Mariner, whose compulsory residence in the Tonga groupe, and whose intimate acquaintance with the native language and manners, have enabled him to throw more light on the customs, religion, character and present state of the inhabitants of those islands and the kindred tribes spread over the South Sea, than any preceding writer. Relative to the Sandwich Islands, much important information has been obtained from the unfortunate Archibald Campbell's *Voyage round the World.* The circumnavigation of Captain Kotzebue, in the Russian ship Rurick, has furnished us with particulars of the Radack chain, which appear to have been first discovered by that officer; and for great part of our account of New Zealand we are indebted to the *Journal of a Ten Month's Residence* there by Captain Cruise of the 84th regiment. The particulars of the astonishing work of civilization, which, commencing at Otaheite and the neighbouring islands, has reached the Sandwich groupe, and is extending to other parts of the Great Ocean, we owe to the Annual Reports of the London Missionary Society, to whose unwearied zeal in the propagation of Christianity this glorious result is chiefly to be ascribed.

To Lieutenant Gardiner of the Royal Navy, we have also to express our best acknowledgments for the obliging communication of original drawings, from which we have selected some of the figures representing natives of the Marquesas. Many of the other engravings which embellish these volumes are copied from the interesting designs of M. Choris, who, as draughtsman, accompanied Captain Kotzebue in his late voyage of discovery.

CONTENTS.

VOL. I.

VOL. II.

Directions for placing the Plates.

VOL. I.

VOL. II.

SOUTH SEA ISLANDS

In Miniature.

INTRODUCTION.

THE concurrent testimony of modern navigators demonstrates that we must look to the extensive countries situated to the west of the great ocean, which we call indiscriminately the Pacific Ocean or South Sea, for the origin of the inhabitants of the innumerable Islands by which it is studded. Their languages, arts, manners and customs, prove their derivation from an Asiatic stock. The domestic animals and plants which have every where accompanied them, and all of which belong to the Old World, proclaim the regions from which they were brought.

We find the sugar-cane, the plantain, the paper-mulberry tree, the banana, yam and sweet potatoe; and of animals, the domestic fowl, diffused over the whole ocean as far as Easter Island; the bread-fruit and some other vegetable productions, the hog and the dog, as far as the Society, Marquesas and Sandwich Islands. New Zealand possessed the latter only, and the Society Islands only the former, though the dog was known there by name. Neither of these animals is found in the Carolines and other islands eastward of the Philippines, between the equator and the northern tropic.

In some of the Asiatic islands, we find the natives dressed in the same sort of stuff made of the inner bark of trees, which is universally worn by the people of the South Sea: and the same kind of personal ornaments are common to both.

It would be in vain to attempt to trace back to one principle and one source the various modifications of the prohibitory laws and customs of the *taboo*, which separate the sexes, raise insuperable barriers between the different classes of the people, and constitute the basis of social order among all the tribes of the South Sea. The aid of written records is wanting to enable us to deduce them from the religious and civil system of other known nations. It is impossible, however, not to recognize in these and other usages, such as the prevailing practice of circumcision and the offering of first-fruits, much of the spirit of the Mosaic legislation, undoubtedly conveyed to these people through the channel of the institutions of Mahomet. In their treatment of the female sex, in their constant ablutions, in their notions respecting the contact of a dead body, in their ideas of paradise, and in some of their religious ceremonies, the intelligent reader will not fail to discover also a strong resemblance to the faith and practice inculcated by the Prophet.

The inequality of classes, founded on arrangements which nearly resemble the feudal system of our ancestors, and which are held most sacred by the islanders of the South Sea; and the devotion of certain classes in some of these tribes, as in the Tonga islands, to particular professions and modes of life, remind us of India. The voluntary death of the wife in the Feejee islands, and the similar custom which prevailed in the family of Tooitonga, in the Tonga islands, likewise refer us to the same country. The use of the beverage called *cava*, *ava*, and *yava*, is found to be almost universal among the South Sea islanders. Would not this appellation seem to indicate some connexion with the Asiatic nations, who are equally addicted to the drinking of *cave* or *cahveh*, called by us coffee: but which term the Orientals apply also to wine or any intoxicating liquor? Shaw, indeed, tells us that the word is of Persian origin, and signifies drinking or any thing fit to drink. The coincidence at least is striking. We leave to those who are better qualified for such researches to determine what weight is due to this conjecture.

Should it now be asked how and at what time a people of Asiatic origin could have dispersed itself over the remotest islands of this vast ocean, in a direction contrary to the course of the wind, bringing along with it the domestic animals and useful plants of its native land; how the different tribes in their separation from each other could there preserve the same manners and the same arts, and notwithstanding the want of writing and the habit of arbitrary innovations, yet speak but one common language: we can only confess our ignorance on these points.

A gradual extension from west to east, occasioned by accidental causes, is the hypothesis which seems to present the fewest difficulties. Facts recorded by modern navigators tend to prove that the currents set in that direction though the winds blow the contrary way. Kotzebue's recent voyage furnishes two remarkable illustrations of this point, which may serve to give some notion of the manner in which many of those islands were originally planted as well as peopled. Having occasion to mention Captain Alexander Adams, commander of a vessel belonging to the king of the Sandwich Islands, Kotzebue gives the following curious extract from the journal kept by that officer while mate of the British brig Forester.

Brig Forester, the 24th March, 1815, *at Sea near the coast of California, lat.* 30° 45′ *north, long.* 233° 3′ *east.*

During a strong wind from w. n. w. and rainy weather, we descried in the morning at 6 o'clock, a ship at a small distance, the appearance of which convinced us that she stood in need of assistance. We immediately directed our course towards her, and recognized her to be a Japanese vessel which had lost her masts and rudder. I was sent on board by the Captain, and found in the ship only three dying Japanese, the captain and two sailors. I instantly had the unfortunate men carried to our brig, where they were perfectly recovered after four months careful attendance. We learned from these people that they came from the port of Osaka, and were bound to another commercial town, but had been overtaken immediately on their departure by a storm, and lost their mast and rudder. They had been, up to this day, the sport of the waves for *seventeen* months, and out of a crew of thirty-five men, these three only survived, and they too must soon have perished of hunger.

This note, observes Kotzebue, is so far remarkable, as it proves that the currents in these seas, that is, north of the tropics, always keep their direction from west to east.

Another passage, however, in the work of the same navigator, shows that this direction is not confined to the north of the tropics, for we are told that Kadoo recognized among the seeds thrown on the shore of the Sandwich islands and Radack, many belonging to native plants of the Carolines.

We have already adverted to the identity of the mother language of the different dialects spoken by the various tribes scattered over so immense a space. This is believed to be the Malay. The researches of modern navigators during the last fifty years, have shown that New Zealand and all the intermediate archipelagoes as far as Easter Island, as well as the detached groupe of the Sandwich Islands, are inhabited by descendants from the same stock, who have arrived almost every where at the sam e degree of civilization, and speak the same language, the dialects of which are distinguished only by local differences in pronunciation; so that mariners who have collected words on one island have often been able with their help to make themselves understood in others very remote. Thus Tupia, a native of the Society Islands, could render himself intelligible to the New Zealanders, and the people of the Sandwich Islands find no difficulty to understand those of Otaheite.

Whatever difference there is between them seems owing to arbitrary changes. In Otaheite, at the accession of a sovereign, or on any other similar occasions, words belonging to the general language are banished and replaced by new ones. This strange custom, according to the account of a Spaniard named Marini, settled in the Sandwich Islands, is not unknown there. He informed the Russian navigators, that about the year 1800, the late king Tammeamea, on occasion of the birth of a son, invented and began to introduce a new language. The new words were neither related to, nor derived from any roots of the current language. Some powerful chiefs, displeased at the change, are said to have destroyed by poison the child who caused it: at his death the attempt was relinquished, and the old language again adopted.

It appears, that the language of religion employed in these islands is a peculiar one, differing from the spoken language, and not understood by the common people. This language does not appear to have received that attention which it deserves from Europeans who have had opportunities of residing in the islands of the South Sea. It might be the ancient, uncorrupted language of the people, and in this case, the careful study of it could not fail to assist essentially in solving the problem of their origin.

In the inhabitants of the Carolines and the islands situated to the east of the Philippines, we find a second family of nations who are variously connected by the same manners and arts, and by great skill in navigation and trade. They are a peaceable, and, in most respects, amiable people, who worship no idols, live without domestic animals on the productions of the earth, and offer its first fruits to invisible deities. They build the most ingenious vessels, and perform distant voyages by means of their accurate knowledge of the stars, currents and monsoons.

Notwithstanding the great resemblance of most of these tribes, and their frequent intercourse with each other, a great diversity of languages prevails among them. The dialect of the Pelew Islands seems to differ most from the common language of the eastern South Sea Islanders, and that of Radack to approach the nearest to it.

Lastly, the inhabitants of the New Hebrides and New Caledonia belong to a third race, the Papuas or Austral Negroes, who are spread also over New Holland, New Guinea and the groupe of islands situated to the eastward of the latter. The languages of the tribes of this third race, are not only totally different from the general language of the South Sea Islanders, but even entirely dissimilar to one another.

We shall commence our survey of the first of these great families, with the Friendly Islanders, concerning whom we possess more authentic and circumstantial information than we have yet received respecting any of their kindred tribes. For this we are indebted to the valuable work, compiled by Dr. Martin from the materials furnished by Mr. William Mariner, an intelligent Englishman, who, during a residence of some years in the Tonga groupe, enjoyed peculiarly favourable opportunities of obtaining a most intimate acquaintance with the language, religion and customs of the natives. Much that is related concerning them will equally apply to the people of other islands of which we shall have occasion to treat: this part will therefore serve to illustrate the succeeding portions of these volumes, and a reference to it will probably enable us to save many repetitions.

THE FRIENDLY ISLANDS.

This cluster of islands, situated between the 18th and 22d degree of south latitude, and between the 175th and 176th west longitude from Greenwich, was thus named by Captain Cook on account of the friendship which appeared to subsist among the inhabitants and their courteous behavour to strangers. Tasman, a Dutch navigator, by whom they were visited about the middle of the seventeenth century, gave to the three principal islands the names of New Amsterdam, Rotterdam and Middleburg, which bear the native appellations of Tonga, Anamooka, and Vavaoo, Lefooga and the cluster called the Hapaee islands complete the groupe. It comprises more than sixty islands small and large. Tonga, the most extensive, is about twenty-one miles in length and thirteen broad; it has always been considered by the natives as the most noble of all the Friendly Islands, and from time immemorial, the greatest chiefs have been accustomed to make it their principal place of residence, and to be buried there after their decease in the tombs of their ancestors. It gives name also by way of pre-eminence to all these islands taken collectively, and is distinguished by the epithet of *taboo* or sacred, whence it has been erroneously denominated by navigators and in our charts, Tongataboo.

The natives of these islands seem to be descended from one common stock with those of other tribes scattered over the Pacific Ocean, from most of whom, however, according to the picture of them drawn by Mr. Mariner, they are advantageously distinguished.

They seldom exceed the common stature, but are very strong and well made, especially as to their limbs. They are generally broad about the shoulders; and though the muscular disposition of the men, which seems to be a consequence of much action, rather conveys the appearance of strength than of beauty, still many of them are really handsome.

The women are not so much distinguished from the men by their features as by their general form, which is mostly well proportioned. The features of many are so delicate as not only to be a true index of their sex, but to lay claim to a considerable share of beauty and expression; while some are absolutely perfect models of a beautiful figure. But the most remarkable distinction in the women is the uncommon smallness and delicacy of their fingers, which may be put in competition with the finest in Europe. The general colour is a cast deeper than the copper brown; but several of both sexes have a true olive complexion, and some of the females are even a great deal fairer.

Their hair is in general thick, straight and strong, though a few have it bushy or frizzled. The natural colour is black, but the greatest part of the men and some of the women have it stained of a brown or purple colour, and a few of an orange cast. They wear it cut in various ways. Some cut it off on one side of the head only, others have it entirely cut off excepting a single lock: the women in general wear it short. The men shave the beard by means of the two valves of a species of shell-fish every eight or ten days, and the heads of infants are kept closely shaved with a shark's tooth by the mother.

The operation of tattowing, which will be described at large in another place,* is practised in these islands. It would be deemed unmanly to want this distinction: and all males therefore submit to the operation as soon as they are grown up. They employ a variety of patterns with which they cover the body from above the navel and the thighs nearly to the knees. Women are not subjected to this process; though many choose to have only a few lines or spots thus imprinted on the inside of their hands. The boys at about the age of fourteen undergo a species of circumcision.

* See the Marquesas Islands.

The dress of both sexes is the same. It consists of a piece of cloth, manufactured from the bark of the paper-mulberry tree, called *gnatoo*, or of matting, but mostly the former, about two yards wide and two and a half long; at least so long as to go once and a half round the waist, to which it is confined by a girdle or cord. It is double before, and hangs down like a petticoat to the middle of the leg. The upper part of this garment above the girdle is plaited into several folds, so that when unfolded there is cloth sufficient to draw up and wrap over the head and shoulders in bad weather or at night. There are two or three other ways of wearing this garment, but that here described is considered as the most elegant, and is therefore the most usual among chiefs.

Persons of the inferior class are satisfied with a smaller piece of cloth, and very often wear nothing but a covering made of leaves of plants, or the *muro*, which is a narrow piece of cloth or matting like a sash: this they pass between the thighs and fasten round the waist; but the use of it is chiefly confined to the men.

The women adjust their *gnatoos* in a different way from the other sex; but the principal distinction of their dress is a small mat, about a foot wide, worn round the middle. Old and pregnant women wear the *gnatoo* in front so as to cover the bosom.

The ornaments worn by both sexes are necklaces made of the fruit of the pandanus and various sweet smelling flowers, which go under the general name of *kahulla*. Others are composed of small shells, the wing and leg-bones of birds, and other things, all which hang loose on the breast. Voyagers have erroneously asserted that they also wear strings of teeth round their necks, whereas it is only the forms of teeth cut out of those of the whale, each preserving the shape of a whale's tooth, from an inch to four inches long. In the broadest part is a hole by which they are closely strung and put round the neck, the largest being in front, and the others decreasing in size on each side up to the back of the neck; so that when drawn close, their pointed extremities diverge and form a pleasing ornament on their brown skins, which they highly prize as well on account of its scarcity as its beauty.

It is astonishing with what neatness and how little waste they make these ornaments with such rude instruments, having in general nothing but a common European chisel, or a piece of a saw, or perhaps only a nail flattened and sharpened. Before they procured iron from European ships they employed only a sharp stone for this purpose

With this kind of ivory they likewise inlay their wooden pillows and their clubs.

They adorn their fingers with rings of tortoise-shell, and a number of these joined together are worn as bracelets at the wrists. The lobes of both ears, though most frequently but one, are sometimes perforated with two holes, in which they insert cylindrical pieces of ivory about three inches long.

These people pay strict attention to personal cleanliness. Rising with the dawn, they wrap their *gnatoos* round them and go out to bathe either in the sea or in a pond; or if neither be at hand, they have water poured over them out of cocoa-nut shells. When bathing they wear an apron either of *gnatoo* or of the leaves of the *chi* tree. On their return home, they oil themselves all over, generally with oil scented with the aroma of flowers; great chiefs frequently use the oil of sandal-wood, which they bring from the Feejee islands for the purpose. After this operation they put on their dress with all possible neatness. When they have no evening dances, they again bathe and oil themselves before they retire to rest, which they generally do about sun-set.

The houses of persons of the lower class are small mean huts, but those of the better sort are larger and more comfortable. The dimensions of one of a middling size are about thirty feet long, twenty broad and twelve high. The general form of the houses is oblong, rather approaching to an oval, the ends being closed and the front and back open. The sloping thatched roof, which is generally supported by four posts, in larger houses by six or more, descends to within about four feet of the ground. The chief art in building a house consists in fastening the timbers strongly with plait of different colours, made of the husk of the cocoa-nut, in such a way as to be very ornamental; the colours, which are black, red and yellow, being very tastefully disposed. The thatch of the better sort of houses is made of the dried leaves of the sugar-cane, which will last seven or eight years; and that of common houses of the leaves of the cocoa-nut tree, which require to be renewed every two or three years. The ground within, being raised about a foot, is beaten down hard, and covered with leaves of the cocoa-nut tree, dried grass, or leaves of the *ifi* tree, and over these is laid a bleached matting, made of the young leaves of the cocoa-nut tree. The house which forms in reality but one apartment, is occasionally divided by screens of matting six or eight feet high. In rainy weather, or at night when it is cold, a sort of blind, attached to the eaves of the open sides of the house, can be let down. These blinds are made of long mats, about six inches wide, one above another, and rather overlapping; and they are so contrived as to draw up by means of strings like our Venetian blinds. Common houses have not these blinds, but a few mats are hung up instead of them whenever occasion requires.

In every island there is a number of very extensive buildings, at which public assemblies, festivities, and ceremonies, are held; Vavaoo, being a large island, has fourteen or fifteen: each of these edifices stands in the centre of a grass-plot of about three acres. A place of this kind is called a *marly*.

Their whole furniture consists of a bowl or two for holding a beverage called *cava*, a few gourds, cocoa-nut shells and low wooden stools, which serve them for pillows. These are about two feet long, but only four or five inches high and nearly four broad, bending down in the middle, with four strong legs and circular feet; the whole made of one piece of black or brown wood, neatly polished and inlaid with bits of ivory.

These islanders display great ingenuity in the building, as well as in the navigating of their canoes. The only tools which they used for constructing them, before these seas began to be frequented by Europeans, were hatchets, or rather thick adzes of smooth black stone, augres made of sharks' teeth, fixed on small handles, and rasps of the rough skin of a fish, fastened on flat pieces of wood which also have handles. The cordage is made of the fibres of the cocoa-nut husk, which, though not more than nine or ten inches long, they plait of the thickness of a small quill to any length they please. Their line for fishing is as strong and even as the best of ours, and nearly resembles it in every respect; and they make hooks, both large and small, of pearl shell.

The canoes of the Friendly Islands surpass in neatness and workmanship those of all the neighbouring tribes. They are built of several pieces sewed together like the timbers of their houses with plait, in so neat a manner that on the outside it is difficult to see the joints: all the fastenings are in the inside. The single canoes are from twenty to thirty feet long, and from twenty to twenty-two inches broad in the middle: the stern terminates in a point, and the head is something like the point of a wedge. All these single canoes have outriggers and are sometimes navigated with sails, but more generally with paddles. The two vessels which compose the double canoes, are each about sixty or seventy feet long and four or five broad in the middle: both ends terminate nearly in a point. Two such canoes are fastened together, parallel with each other, about six or seven feet asunder, by strong cross-beams, over which is laid a boarded platform. All the parts composing the double canoe are made as strong and light as the nature of the work will admit: so that they may be immerged in water to the very platform, without being in danger of filling; nor is it possible under any circumstance for them to sink so long as they hold together. Thus they are adapted not only for vessels of burden, but also for distant navigation; and the natives perform voyages of one or two hundred leagues in them. They are rigged with one mast with steps upon the platform, which can be raised or taken down, and navigated with a latteen sail or triangular ones, extended by a long yard which is a little bent.

The sails are made of mats. On the platform is erected a little shed or hut, to screen the crew from the sun and weather: and the vessel carries a moveable fire-hearth, which is a square, shallow wooden trough filled with stones. A strong rope with large stones fastened to the end serves the purpose of an anchor.

Their weapons are clubs, spears, darts, bows and arrows. The former are of a great variety of shapes, but the most common form is quadrangular. By far the greater number are carved over in many chequered patterns, with great ingenuity, and the whole surface of the plain clubs is as highly polished as if finished by an European artist with the best instruments. Their spears are sometimes plain sharp-pointed sticks, and sometimes barbed: and their bows and arrows are of a peculiar construction. The bow, about the thickness of a man's little finger, forms a slight curve when slack; and the convex part is channelled with a single deep groove in which the bow-string is lodged. In order to bend it, instead of drawing it so as to increase the natural curve, they draw it the contrary way, till it forms a curve on the other side. The arrow is made of reed and pointed with hard wood.

Yams, plantains and cocoa-nuts and bread-fruit, compose the greatest part of their vegetable diet. Of their animal food the chief articles are hogs, fowls and fish, the latter commonly eaten raw: but the otherarticles are commonly dressed by baking. Instead of plates they lay their victuals on green plantain leaves. The women are not excluded from taking their meals with the men.

When a person salutes a superior relation, he kisses the hand of the party; if a very superior relation, he kisses the foot; the superior in return kisses the forehead. Their kiss, however, differs much from ours, consisting only in the pressure of the upper lip and nostrils, so as to have more the appearance of smelling. When two equals salute, each applies his upper lip and nostrils to the forehead of the other; or they apply their lips to the lips of each other, but without any movement. Our kiss they never adopt, not even between the sexes, but term it in ridicule, "the white man's kiss."

It appears from Mr. Mariner's work, that the habits of the people of the Friendly Islands have undergone a great change since the time of Captain Cook's last visit. They knew then but little of war; the only quarrels in which they were previously engaged having been among the inhabitants of the Feejee islands, about 120 leagues to the westward; for, having been in the habit of visiting those islands to procure sandal-wood and other commodities, they occasionally assisted one or other of the warlike parties there against their enemies. The bows and arrows previously employed by the people of Tonga were of a weaker kind, less suitable for war than sport, for shooting rats, birds, and other small animals. From the fierce and warlike people of the Feejee islands, however, they soon learned to construct bows and arrows of a much more formidable kind, and became acquainted with a better form of the spear, and a superior method of holding and throwing that weapon. They also imitated them, by degrees, in the practice of painting their faces and using a peculiar dress in time of war, producing a fierce appearance calculated to strike terror into their enemies. These martial improvements were in progress at the time of Captain Cook's visit, but not in general practice; for having few or no civil dissensions among themselves, the knowledge of these things was almost exclusively confined to a few young chiefs and their adherents who had been at the Feejee Islands. The murder of King Toogoo Ahoo, and the long subsequent struggle which gave to Finow the sovereignty of the Tonga islands, have introduced among the inhabitants much of that warlike spirit which animates the Feejee people, whose maxim is, that war and strife are the noble employments of men, and that ease and pleasure are worthy to be courted only by the weak and effeminate.

It was from the Feejee people also that the Tonga islanders acquired a relish for human flesh as well as for war. A famine, which happened some time afterwards, served to make them more familiar with this unnatural food. On this occasion they waylaid and murdered one another to supply themselves with food; and they relate that four brothers during this time of scarcity invited their aunt to come and partake of a large yam, which, they said, they had secretly procured. The poor woman, pleased with the idea of obtaining a good meal, repaired to the house of her nephews, who presently dispatched her, and she herself furnished them with materials for a repast.

During Mariner's residence on the island, after a battle, some of the chiefs proposed to kill their prisoners lest they should escape, and then roast and eat them. This proposal was readily assented to, especially as a scarcity of provisions then prevailed. Some of the prisoners were accordingly dispatched; their flesh was cut up into small portions, washed with sea-water, wrapped in plantain leaves and roasted under hot stones: two or three were embowelled and baked whole in the same manner as a pig. Our countryman could not be prevailed on to partake of this diet, though pressed by hunger, and though the smell of it when cooked was exceedingly delicious.

A remarkable trait in the character of the people of Tonga is, that they never exult in any feats of bravery which they may have performed, but on the contrary take every opportunity of praising their adversaries and making excuses for their failure. The same principle they carry into their wrestling-matches.

Their notions of true bravery appear to be very correct; and they accuse Europeans of a great deal of vanity and selfishness on this point, and unfortunately with too much justice. These noble sentiments, however, belong almost exclusively to chiefs, matabooles and professed warriors: and are not much found among the lowest class, many of whom will beat a dead man about the head with a club, till they have notched and stained it with blood, and pretend that it was done in battle against a living foe.

Here, as in Japan, and perhaps in no other part of the world, is exhibited the extraordinary anomaly of a spiritual chief, who is deemed greatly superior in rank to the king, though the latter possesses the supreme temporal power. This chief is called Tooitonga, and is acknowledged to be the lineal descendant of one of the chief gods who formerly visited these islands. They have also a second divine chief, called Veachi, supposed to be of similar descent but inferior in rank to Tooitonga.

The *how* or king is an arbitrary monarch, deriving his right to rule partly from hereditary succession and partly from military power. The family of Finow, the present *how*, is probably a distant branch of one of the sacred families just mentioned. His father was the first of the family who elevated himself to royalty by usurpation and the expulsion of the then reigning house. Before that time, the *hows*, as far back as credible traditions extend, which is not more than four, or at most five generations, were all relatives to Tooitonga.

Formerly when the Tonga people were a peaceable nation, Tooitonga and Veachi possessed much more influence than they have of late years enjoyed. It is presumed that even in matters of government their advice was then frequently asked and sometimes taken. The late king would not suffer Tooitonga to give him advice on political matters, insisting that he should remain in peace and quietness at his own side of the island; neither would he permit his daughter, who was the principal wife of that chief, to be strangled at his death and buried with him, according to the custom previously observed on such occasions. At the decease of the late Tooitonga, the present king abolished that high title altogether, and at one bold stroke relieved the people from a heavy tax, by suppressing the ceremony of the *inachi*, which will be hereafter described

The *egi*, nobles or chiefs, are persons related to the family of Tooitonga, Veachi or the *how*: and the nearer any family is related to them, the nobler it is; those related to Tooitonga being nobler than those equally related to Veachi, and those related to Veachi being nobler than those equally related to the *how*. Hence there are many *egis* more noble than the king himself; and to such the king when he meets them must pay the same marks of respect as are usual from an inferior to a superior. Thus, were he to touch the superior chief, or any thing personally belonging to him, as his garments or the mat on which he sleeps, he becomes tabooed, as it is termed, and must not feed himself with his own hands, or if he does, it is at the risk of becoming diseased or suffering some other calamity as a punishment from the gods. From this taboo, however, he can release himself by performing the ceremony of *moë-moë*, which consists in touching with both hands the feet of the superior chief or one equal to him, and pressing them against his stomach.

Sitting down is with them a token of respect, as standing up is with us, perhaps because in this posture a man cannot so readily attack the person in whose presence he is. Were even the king to meet any chief of nobler descent than himself, he would be obliged to sit down on the ground till the other had passed him; which is a mark of reverence that is expected from the meanest peasant to any chief whatsoever. For this reason the king never associates with any chief superior to himself, and always endeavours to avoid meeting them; while they, on their part, equally shun him, lest he should be put to the trouble of sitting down while they passed: for, if any one were to forego this ceremony in the presence of a superior, some calamity from the gods would be expected to overtake him.

In every family, nobility descends by the female line; for where the mother is not a noble, the children are not nobles.

The *matabaoles,* who rank next, are a sort of honourable attendants on chiefs; they are their companions, counsellors and advisers. They see that the orders and wishes of their chiefs are duly executed, and may not improperly be called their ministers. It is their business to make themselves acquainted with all rites and ceremonies, and with the manners, customs and affairs of Tonga. Some of them are adepts at some art or profession, such as canoe-building, or superintending funeral rites: they also make themselves acquainted with traditionary records, and hand them down to their sons. When a mataboole dies, his eldest son, or if he have no son, his next brother becomes a mataboole, and all the sons and brothers of matabooles are *mooas*.

The *mooas* are the next class below matabooles; being either the sons or brothers of matabooles, or their descendants. The mooas have much to do in assisting at public ceremonies, and in sharing out food under the direction of matabooles, and on ordinary occasions they sometimes arrange and direct instead of the latter. Like the matabooles, they form part of the retinue of chiefs, and are more or less respected according to the rank of the latter. Most of them are professors of some art. It is the duty of both matabooles and mooas to attend to the good order of society and to the morals of the younger chiefs who are apt to run into excesses and to oppress the lower orders; in which case they admonish them, and if they pay no attention, they report their conduct to the older chiefs and advise what measures should be taken for their amendment. Most of the mooas are professors of some art, and they are highly respected by the lower class.

The *tooas* constitute the lowest order,

or the bulk of the people. They are all by birth peasants ; but some of them are employed occasionally in tattowing, cooking, club-carving and shaving. Such tooas as are evidently related to, and consequently have a chance of becoming mooas, are respected by those who cannot trace such relationship.

Of the different professions some are hereditary and others not: those which are considered the most honourable are followed only by matabooles and mooas. The two lowest of all, cooks and husbandmen, are such by inheritance, and must follow these necessary employments as the business of their lives, if the chiefs to whose service they are attached command them.

Old persons of both sexes are highly honoured on account of their age and experience; insomuch, that it is one of the first moral and religious duties to reverence the gods, the chiefs and aged persons. Women also have considerable respect paid to them on account of their sex, independently of the rank they may hold as nobles. Hence they are not required to perform any hard labour or very menial work. It seems to be a peculiar trait in the character of the Tonga people, when compared with that of the natives of some other South Sea Islands, and with savage nations in general, that they do not load the weaker sex with the heaviest burdens of life: but from generous motives take upon themselves all those laborious tasks which they think inconsistent with the weakness and delicacy of the softer sex. The consequence of this respect paid to women in the Tonga islands is, that they hold themselves in greater respect and exhibit none of that unbridled licentiousness which is so universal among the females of most of the islands of the South Sea. Their reserve and modesty are strongly attested by Captain Cook, as well as in the more recent work of Mr. Mariner, who assures us, that their behaviour as daughters, wives, and mothers, is far from being unworthy of imitation. They consequently bestow the greatest care on their children, who are never neglected either in regard to personal cleanliness or diet. As they grow older, the boys are made to exercise themselves in athletic sports, and the girls to learn suitable arts and manufactures, and to acquire many little ornamental accomplishments, which tend to render them agreeable companions and objects of esteem—such as plaiting various pretty fanciful devices in flowers, which they present in token of respect to their fathers, brothers, or superior chiefs. The female chiefs are allowed to imitate the men, in having attendants of their own sex, consisting of the wives and daughters of inferior chiefs and matabooles; and such an association tends, as may easily be conceived, to support their rank and independence.

The women of the inferior class are chiefly employed in the manufacture of *gnatoo*, or cloth made of the inner bark of the paper mulberry-tree; also in making mats and baskets of various kinds, an occupation that is followed even by those of some rank both for amusement and profit. In these lighter works and in those offices which are requisite for domestic comfort, health and cleanliness, the women of Tonga are engaged.

Marriage is frequently independent of the consent of the woman, who is betrothed at an early age to some chief, mataboole or mooa: perhaps about one-third have been thus betrothed, and the remaining two-thirds have married with their free consent. Whilst unmarried, a female is mistress of her own conduct; but when once she becomes a wife, it is her positive duty to remain true to her husband, who, for any cause of offence, may dismiss her whenever he pleases. She is then at liberty to marry another; and in this way many a woman has four or five husbands in the course of her life, without the least disparagement to her character.

The ceremonies attending the marriage of Finow's eldest son, who has since succeeded him in the sovereignty, are detailed by Mr. Mariner. This prince, who had been some years absent at Hamoa, or Navigators' Islands, had there married two wives, with whom he returned to Vavaoo. Here he found that two daughters of chiefs had been reserved for him, and he resolved to marry them also, with the ceremonies common in the Navigators' Islands, which however differ but little from those of Tonga.

On the morning of the day of marriage most of the people of the lower class were employing in bringing from different parts of the island yams, ripe plantains and bananas, cocoa-nuts, bread fruit, fish and cakes. These were piled up on the *marly* in four large heaps, with a baked pig on the top of each. The people assembled on the spot, dressed in new garments, adorned with wreaths of flowers, and with red ribands made of the fine membrane of the leaf of a tree much resembling silk. Their bodies were anointed with sweet-scented oil. The spectators seated themselves in two segments of a circle on the right and left hand of the king and his matabooles: at their terminations on either side stood the Hapaee people who had accompanied the prince, provided with clubs made of the green branches of the cocoa-nut tree. The prince likewise armed with a club stood among his Hamoa companions.

The two brides, dressed in the finest

Hamoa mats, and veiled in the finest *gnatoo*, were now conducted from Finow's habitation, to the house on the *marly* and seated on bales of gnatoo. Here their feet, hands, faces and bosoms were anointed with a mixture of sandal-wood oil and the purest turmeric, which gave to their skins a deep orange tint. Here they remained seated to behold the fight which was about to ensue between the inhabitants of Hapaee and their friends from Hamoa.

The parties being ready, the method of challenge is as follows: a man from one side runs over to the opposite party, sits down before it, and asks if any person will engage him. The person who chooses to accept the challenge comes forward brandishing his club, and the two combatants proceed to the middle of the circle, each attended by one from his own party to assist as second. They then decide whether they shall fight after the Tonga or the Hamoa fashion, which differ in this respect, that the Hamoa custom allows a man to beat his antagonist after he is knocked down as long as he perceives signs of motion; whereas the Tonga mode only allows him to flourish his club over his fallen foe, and the fight is at an end. Having settled this point the champions engage, and when they have finished, another party comes on in the same way; though frequently three or four sets of combatants are engaged at one ti me.

When a man gains a victory, his party gives a shout of approbation; the champion then advances towards the chief who presides at the head of the circle, sits down before him in token of respect, then rises immediately and returns to his party. In these club-fights the prince performed great feats of bravery; he fought fourteen or fifteen battles and always proved victorious.

The club-fighting being over, on a signal from Finow the boxing and wrestling-matches commenced. These feats, which will be described in another place, being over, the prince and his chiefs retired to the neighbouring houses to dress their heads with a sort of turban made of white *gnatoo*, adorned with red feathers. Thus equipped they returned to the *marly* ; the chiefs again sat down among their own party, while the prince went to his two brides who were still sitting in the house, and taking one by each hand led them forth upon the *marly* amid the acclamations of the people ; the young chiefs and their companions from Hamoa singing a song in the language of the Navigators' Islands and beating time with their hands. Meanwhile the prince, with a slow and dignified step, led the ladies to the bottom of the *marly* and then returned, amid the same acclamations into the house and conducted them back to their seat on the bales of *gnatoo*. He then commenced a dance on the *marly* with the young chiefs who had put on turbans. During this dance, which was also after the manner of Hamoa, the brides were led to the habitation of the bridegroom; and when it was over, the provisions were shared out. The three larger portions were allotted to the strangers from other islands; and at an appointed signal a general scramble commenced for the other. The baked pig on the top of the heap was soon brought down and mauled in a most miserable way, being torn piecemeal and so covered with sand and dirt as to be quite uneatable.

The ceremony concluded with a general boxing-match, the men from the north of the island combating those from the south, till the former at length drove their opponents completely off the ground. At such fights the combatants are allowed to wear turbans, to produce a more accurate representation of a real fight: for turbans are not permitted to be worn but in time of war, and then only by those who are going to battle, unless on occasion of formal rejoicings, or at night by chiefs and matabooles, or by the common people when at work in the fields or in canoes. At all other times it would be deemed disrespectful to wear a head-dress; for though no chief be present, yet some god may be at hand unseen. This custom is kept up with such strictness, that if a man were to wear a head-dress on other occasions, he would be sure to be knocked down by the first person he met who was his superior and perhaps even only his equal.

In the evening the large house on the *marly* was lighted up with flambeaux; singers and dancers assembled, and on the arrival of the prince, began a vocal concert which lasted four or five hours, and during which the prince, not being much interested in it, gave them the slip.

It is customary in these islands for women of superior rank to act the part of mothers to children or grown-up young persons who are not their own, for the purpose of providing them or causing them to be provided with all the conveniencies of life; and this is often done though the natural mothers be living and residing near the spot. Thus one of the wives of Finow, the father of the present king, was appointed by her husband to be foster-mother to Mr. Mariner; and to her he was indebted not only for many comforts and conveniencies which he would not otherwise have enjoyed, but also for his intimate knowledge of the language and customs of Tonga, in which she took great pains to instruct him.

Various facts are recorded in Mariner's work which place the character of the women in a very favourable light. This traveller relates that during his residence on the islands, the king had connived at the death of his brother, Toobo Neuha, chief of Vavaoo, and conferred his authority on his aunt. The latter soon afterwards assembled the principle persons of the island, acknowledged the obligations she owed to her murdered nephew, and proposed that they should liberate themselves from the yoke of Finow, his brother. This proposal led to a long discussion, till at length an old woman, sister to the female chief, rushed into the assembly, armed with a club and spear, and brandishing them, asked with a loud voice, why they hesitated so long in an affair in which honour clearly pointed out the only proper path to pursue.

"But," added she, "if the men are turned women, the women shall turn men, and revenge the death of their murdered chief! Let then the men stand idly looking on, and when we women are sacrificed in the glorious cause, the example may perhaps excite them to fight and die in the same spirited endeavour to support and defend their rights!" This warlike declaration of the heroine had the desired effect upon the chiefs of the other sex.

Other instances of the like resolution and heroism in the sex might be adduced from the same writer.

According to the religion of the Tonga islanders, there are *hotooas*, gods or superior beings, who possess the power of dispensing good and evil to mankind according to their merits. Another class of hotooas, having the like power, but in an inferior degree, consists of the souls of all deceased chiefs and mata-booles. Besides these, there are gods who inflict petty troubles not as a punishment, but indiscriminately and from a purely mischievous disposition. All other evils are dispensed by the gods on account of the neglect of some religious duty, either in the person of the sufferers, or the chiefs whom they serve; and the contrary of good. The *tooas*, or lower class of the people, have no souls, or such only as dissolve with the body after death, which of course terminates their existence.

The Tonga people universally and positively believe that the place of residence of their gods and of the souls of their nobles and matabooles is a large island, situated at a great distance to the north-west of their own islands, and called Bolotoo. It is supposed to be much more extensive than all their islands put together, to be well stocked with all kinds of useful and ornamental plants, always covered with the most beautiful flowers and the richest fruits, which impregnate the atmosphere with their fragrance, and if plucked are immediately replaced by others. The island also abounds in the most beautiful birds of all kinds, and in hogs, all which are immortal, unless killed to feast the hotooas, when new ones instantly come into existence to supply their place. Bolotoo is supposed to be so far off that it is dangerous for canoes to attempt going thither, and they moreover believe that they should not find it unless it happened to be the particular pleasure of the gods.

It is related, however, that a long time ago, a Tonga canoe, on her return from the Feejee islands, was driven by contrary winds to Bolotoo. The crew, not knowing where they were, landed; being in want of provisions, and seeing that the country abounded in all sorts of fruit, they attempted to pluck some bread-fruit, but to their inexpressible astonishment, they could no more lay hold of it than of a shadow. They walked through the trunks of the trees, and passed through the substance of the houses, which were built like those of Tonga, without feeling any resistance. At length they saw some of the *hotooas*, who exhorted them to depart immediately, as they had no food suitable for them, and promised a fair wind and a speedy passage. They complied, and instantly put to sea: after sailing two days with incredible velocity, they arrived at Hamoa, where they wished to touch, and thence returned to Tonga. In a few days they all died, as a natural consequence of having been at Bolotoo, the air of which infects mortal bodies with speedy death.

The primitive *hotooas* as well as the souls of the nobles, are believed to possess the power of coming back to Tonga to inspire priests and others, or to appear visibly to them: but they are transported thither by mere volition. The first class are supposed to amount to about three hundred, but the names of very few are known, and that only to some chiefs and matabooles. To about twenty of the primitive hotooas are dedicated houses, built in the usual style: but more care is generally taken both in their construction and to keep them in good order and decorate their inclosures with flowers. The souls of *egis* or nobles have no houses dedicated to them, but are invoked at their graves, which are held as sacred as the temples.

The mischievous gods, who are invisible, reside more commonly at Tonga than at Bolotoo. They have neither houses nor priests. Only five or six of these are supposed to be particularly active. They lead travellers astray, trip them up, pinch them, jump on their backs in the dark, and cause the nightmare and frightful dreams.

Besides these divinities, there is a god who lies prostrate and supports the earth. He is of enormous size, and surpasses all the others in bulk. He never quits his situation, of course never inspires any one, and has no house dedicated to him. When he finds himself uneasy and endeavours to turn, this is the cause of earthquakes: on such occasions the people shout aloud and beat the ground with sticks, to make him lie still. They have no idea of what he lies on, nor ever think of inquiring about it, saying it would be folly to do so, for who can go thither and see?

As to the origin of the earth, they believe that there was in the beginning no land above the ocean but the island of Bolotoo, which, like the gods, the heavenly bodies and the sea, has probably always existed. One day, Tangaloa, the god of arts and inventions, went out to fish in the Great Ocean, and having let down his hook and line from the sky into the sea, he suddenly felt a powerful resistance. In the belief that he had caught an immense fish, he exerted his strength, and several points of rocks presently appeared above the surface: these increased in number and extent the more he drew in his line. It was now evident that his hook had caught in the rocky bottom of the ocean, which was thus fast rising to form one immense continent, when unluckily the line broke and the Tonga islands alone remain to prove the failure of the attempt. The rock in which the hook was fixed, and the hole where it caught, are shewn to this day in the island of Hoonga. The hook itself was in the possession of the Tooitonga family till between thirty and forty years ago, when it was accidentally burned together with the house in which it was kept.

Through the influence of Tangaloa and the other gods, the land which he had thus brought to light was soon stocked with all kinds of trees, herbs, and animals, resembling those of Bolotoo, but subject to decay and death. Wishing that Tonga should be inhabited by rational beings also, he said to his two sons: "Go, take with you your wives and dwell in the world at Tonga; divide the land into two portions, and dwell separate from each other." They departed accordingly. The name of the elder was Toobo, and that of the younger Vaca-acow-ooli. The latter was a very wise young man; for it was he who first made axes, beads, cloth and looking-glasses. Toobo, on the contrary, was extremely indolent, sauntering about and sleeping, and envying exceedingly the works of his brother. Tired at length of begging his goods, he bethought himself to kill him, but concealed his wicked intention: accordingly meeting his brother walking, he struck him till he was dead. Their father, in very great anger, presently came from Bolotoo, and asked him: "Why hast thou slain thy brother? Couldst thou not work like him? O thou wicked one, begone! Go with my commands to the family of Vaca-acow-ooli; tell them to come hither." When they had come, Tangaloa straightway ordered them thus: "Put your canoes to sea, and sail to the great land which is to the east, and take up your abode there. Be your skins white like your minds, for your minds are pure; you shall be wise, making axes and all sorts of riches, and shall have large canoes. I will go myself and command the wind to blow from your land to Tonga, but the Tonga people shall not be able to go to you with their bad canoes." Tangaloa then spoke thus to the others: "Ye shall be black, because your minds are bad; ye shall not be wise in useful things, neither shall ye go to the great land of your brothers; how can ye go with your bad canoes? But your brothers shall come to Tonga and trade with you as they please."

The reader will not fail to perceive the resemblance between this tradition and our scriptural history of Cain and Abel. The oldest men assured Mr. Mariner that it was an ancient traditionary record among them and founded in truth; whence it is not improbable that the ancestors of these people came from a country to which either Judaism or Christianity had extended itself.

They have several other accounts of the origin of mankind, or rather of the first inhabitants of Tonga; but most of them extremely vague and ridiculous, probably invented for the purpose of passing the time: the natives of these islands being very fond of inventing tales for amusement, like the continental nations west of them, but deficient in the poetical elegance of the latter. The account more generally known and believed than the preceding, and in which we may still trace some coincidence with the scriptural history of the first pair in Paradise, is as follows:—

At a time when the Tonga islands were already in existence, but not yet peopled with intelligent beings, some of the inferior gods of Bolotoo, being desirous to see the new world which Tangaloa had fished up, put to sea to the number of about two hundred, male and female, in a large canoe, and arrived at the island of Tonga. They were so well pleased with the place that they resolved to remain there, and accordingly broke up their canoe to make small ones of it. In a few days, however, two or three of them died. This phenomenon alarmed all the rest; for decay and death were totally inconsistent with their notions of their immortality. About this time one of them felt himself strangely affected and hence he knew that one of the superior gods was coming from Bolotoo to inspire him: in a short time he was actually inspired, and informed his companions that the chief gods had decreed that, as they had come to Tonga and breathed its air and fed upon its productions, they should become mortal and people the world with mortal beings, and all about them should be subject to decay and death. This denunciation grieved them all exceedingly: they were sorry that they had broken up their canoe, but made another, and some of them put to sea in hopes of regaining Bolotoo, promising if successful to return and fetch their companions, but they sought the land of the gods in vain, and were obliged to go back sorely afflicted to Tonga.

The priests like the Shamans of the nations of the interior of Asia, are supposed on certain occasions to be inspired by the gods and to make known their will. When a chief purposes to consult the oracles, he causes a hog to be killed and a basket of yams and two ripe plantains to be procured overnight. These articles are carried in the morning either to the residence of the priest or to the place where he may happen to be. Hither also the chiefs and matabooles repair clothed in mats, and seat themselves in such a manner as would form a circle or ellipsis about the priest if a considerable vacant space were not left opposite to the latter. In this space sits a man who prepares cava.

As soon as all are seated, the priest is supposed to be inspired by the god. He sits for some time in silence, his hands clasped, his eyes cast down and perfectly still. The cava is meanwhile prepared and the provisions are shared out.

The mataboo les begin to consult him; sometimes he answers them, at others not: frequently not returning a word till the repast is finished. When he speaks, he generally begins in a low and very altered tone of voice, which gradually rises to near its natural pitch though sometimes a little above it. All that he says is supposed to be the declaration of the god, and he delivers himself in general without any apparent inward emotion or outward agitation. Sometimes, however, his countenance becomes fierce and inflamed and his whole frame agitated; he is seized with a universal trembling; the perspiration covers his forehead; his lips turn black and are convulsed; at length his bosom heaves, floods of tears start from his eyes and his utterance is choked. Before this paroxysm comes on and after it has gradually subsided, he often eats as much as would satisfy four hungry men. The fit being now gone off, he remains calm for some time, then takes up a club that is placed by him for the purpose, turns it over and views it attentively; he then looks up earnestly first to the right, then to the left and then again at the club, and so on for several times. At length he suddenly raises the club and after a momentary pause, strikes the ground or the nearest part of the house with considerable force; the god immediately leaves him, and he rises and retires to the back of the ring among the people. Mr. Mariner, who enjoyed the best opportunities for observation, has no reason to suspect this violent perturbation of the priests to be feigned for the purpose of popular deception

Persons who are not priests, particularly females, sometimes imagine themselves to be visited by gods, but they are never affected in the manner described above. They are generally low-spirited and thoughtful; as the symptoms encrease, they shed a profusion of tears: and sometimes they swoon away for a few minutes; the height of the paroxysm generally lasts from a quarter to half an hour. These also are called fits of inspiration, and are firmly believed to be visitations from some god, who accuses the party of neglect of religious duty, not by an apparent audible warning, but by an inward compunction of conscience.

These imaginations sometimes produce very serious consequences, of which Mariner gives the following instance. A chief, a very handsome young man, became inspired but did not know by whom. He suddenly felt himself extremely low-spirited and presently swooned away; on coming to himself, being still very ill, he was taken, as is usual in such cases, to the house of a priest, who informed the sick chief that a woman, whose name he mentioned, who had died two years before and was then in Bolotoo, had inspired him; that she was deeply in love with him and wished him to die, (which event would happen in a few days) that she might have him near her. The chief replied, that he had seen the figure of a female two or three successive nights in his sleep, and had begun to suspect that he was inspired by her, though he knew not who she was. He died two days afterwards. Mr. Mariner visited the sick chief three or four times at the house of the priest and heard the latter predict his death and assign its cause.

The term *taboo* implies a system of religious or rather superstitious prohibitions, the influence of which extends over almost all the islands of the South Seas. It is not only applied to the thing prohibited but to the prohibition itself, and frequently in sacred matters to the transgressor of the prohibition. A piece of ground consecrated to a god either by express declaration or by the burial of a great chief is said to be taboo. Persons fighting on consecrated ground would also be considered as taboo, and a man in this same predicament would have to make some sacrifice to the gods as an atonement for the sacrilege.

If any one touches a superior chief or superior relation, or any thing immediately belonging to him, he becomes tabooed and must not feed himself with his own hands till he has removed this taboo, which is done by the performance of a ceremony called *moĕ-moë*. It consists in touching the soles of the feet of any superior chief, first with the palm and then the back of each hand; after which the hands must be rinsed in a little water, or if there is no water near, they may be rubbed with any part of the stem of the plantain or banana-tree, the moisture of which will answer the same purpose. He may then feed himself without danger of any disease which he would as they think otherwise incur by eating with tabooed hands. Should any one believe, however, that he may have already eaten unknowingly with tabooed hands, he then sits down before a chief, and taking up his foot, presses the sole of it against his own abdomen, that the food may do him no injury, and that he may not swell up and die.

It is tabooed also to eat when a superior relation is present, unless the back is turned towards him, and to eat food which a superior relation or chief has touched; and if either of these taboos is accidentally violated, the ceremony just described must be performed.

No person can touch a dead chief without being tabooed for ten months, excepting chiefs who are only tabooed for three, four, or five months according to the superiority of the deceased; again excepting the case of Tooitonga, when even the greatest chief would be tabooed ten months. During the time that a man is tabooed, he must not use his own hands to feed himself, but must be fed by another: he must not even use a tooth-pick himself but must guide another person's hand who holds it. If he is hungry and there is none to feed him, he must go down on his hands and knees and pick up his victuals with his mouth; and if he infringes any of these rules, it is firmly believed that he will swell up and die. The natives frequently saw Mariner feed himself with his hands after touching the dead bodies of chiefs, without observing any decline in his health; but this they attributed to his being a foreigner and governed by different gods.

Certain kinds of food, as turtle and certain species of fish, from something in their nature are tabooed, and must not be eaten until a small portion has been set apart for the gods. Any other kind of food may be rendered taboo by a prohibition laid on it. Fruits and flowers when tabooed are generally marked to be so by pieces of white *tupa* (which is gnatoo not printed or stamped with any pattern,) or a piece of plaitin the shape of a lizard or shark. Cava cannot be tabooed by the touch of any chief of what rank soever.

There is no public religious rite and scarcely any private ceremony of which the drinking of *cava* does not form an important or at least a usual part. This liquor is made with the root of a plant of the same name, which belongs to the pepper species, and is employed for the same purpose in most of the islands of the South Sea. At all cava parties provisions also are dealt out. When the party is very large, it is held for the sake of room on a *marly'*, the chief who presides sitting within the eaves of the house, while the rest of the company form a circle from either hand. The root being scraped clean is cut into small pieces which are well chewed, but only by young persons who have good teeth and clean mouths, and then deposited in a wooden bowl, which for a numerous party may be three feet in diameter and one deep: water is added; this is poured out of cocoa-nut shells till there is sufficient. The root being well mixed with the water, a large quantity of the fibrous bark of a tree, resembling our willow-shavings, is put into the bowl; with this the operator collects the dregs, and wrings the liquid from them. This process is repeated till the cava is sufficiently clear for drinking. It is then served out in cups according to rank, together with the provisions, which generally consist of yams, bananas or plantains, and sometimes fowls, or a baked pig. Every fresh bowl is supplied with a fresh quantity of provisions.

In religious ceremonies the cava-party is held mostly in the morning, before the house consecrated to the god who is to be consulted; a priest presides in this case and has always the first cup. Women never attend the cava-parties of the men, but have similar parties to themselves, which do not last so long, and at which the company drink more moderately than the other sex.

Like all unenlightened people, the natives of Tonga are firm believers in the influence of dreams, omens and charms. Thunder and lightning, which are supposed to be occasioned by the discussious of the gods in Bolotoo, betoken public calamities and sneezing is deemed a sure indication of the failure of any undertaking. A species of bird which they call *chicota*, is apt to make a sudden descent and to dart close by a person, making a shrieking noise. This bird is thought to possess a knowledge of futurity and to give warning by this action of some impending evil.

The principal of the charms practised here is called *tatao*. It consists in hiding upon the grave or in any part of the *fytoca* or house erected over it, some portion of the wearing apparel of an inferior relation of the deceased, in consequence of which that relation will sicken or die: or it may be buried in the house consecrated to the tutelar god of the family. When the grave of a deceased person is made use of, this charm is believed not to have the desired effect unless the deceased was of superior rank to the person on whom the charm is practised.

Another charm, the object of which is to discover whether a sick person will recover or not, is practised by spinning a cocoa-nut with the husk on. For this purpose the nut being placed on the ground, a relation of the sick person determines that if the nut when again at rest after being spun points to a certain quarter, for instance the east, the patient will recover. He then prays aloud to the tutelar god of the family, that he would be pleased to direct the nut so that it may indicate the truth: the nut being then spun, the result is received with perfect confidence as declaratory of the intentions of the gods.

The curses of these rude people would, from their variety and energy, shame the comparatively meagre vocabulary of our most accomplished reprobates. It is only when a number of these curses is pronounced in a string, firmly and with real malevolence, that they are supposed to have any effect, and not even then if the party cursing be considerably lower in rank than the person to whom they are addressed. As a sample of these maledictions, of which forty or even double that number are sometimes uttered at once, the following may suffice:—Dig up your father by moonlight and make soup of his bones; bake his skin to cracknel; gnaw his skull; devour your mother; chew the heart of your grandfather; swallow the eyes of your uncle; strike your god; eat the gristly bones of your children; suck out the brains of your grandmother, &c.

The opinion which these people entertain of writing and of books, may be gathered from the following statement by Mariner.

As he had in his possession on his first arrival some books and writing paper, Finow, the king, who strongly befriended him, often found him reading and writing. One day he desired him to give up all his books and papers which he ordered to be burnt. On requiring an explanation of this extraordinary conduct, he learned that the king would not allow him to practise witchcraft to the prejudice of his people; for it was well known to the king and many others, that the books and papers were instruments and means of invocation, to bring some evil or plague upon the country. After Mr. Mariner had learned their language, the king explained to him his opinion on this subject, and gave his reason for it in the following narrative:

Some years ago, on the arrival of a European vessel, one of the white men, named Morgan, came to live among them by choice. For a considerable time he lived on friendly terms with the natives and was much respected by them. Some time afterwards, there came another European vessel, (the Duff, Captain Wilson, with the missionaries,) and from this ship also several white men came by choice to live among them. These last built a house in which they would frequently shut themselves up, to sing and perform ceremonies, as Finow expressed it

For some time things went on well; a quarrel at length took place between Morgan and the other white men, originally about an iron pot which he wanted to borrow of them, and afterwards about some pigs which they accused him of stealing. Hereupon they informed the chiefs that Morgan had been a bad man in his own country and was under sentence of banishment from it for his crimes, but from the full execution of which he had escaped. (He had actually escaped from Botany Bay.) The people then began to insult Morgan in every possible way, so that he was very uncomfortable and his life often in great danger. Morgan in his turn told the chiefs who his countrymen were; he declared that they were men sent out by the king of England to bring a pestilence on the people of Tonga, and that they accordingly shut themselves up in their house to perform witchcraft and incantations, which were the cause of the pestilence that then raged, and that all their books were books of witchcraft. (There was actually at the time an epidemic disease which was very fatal to the chiefs, two or three of whom died every day). The chiefs began to be alarmed by Morgan's statement; there certainly was a great mortality among them; the white men often assembled and sang very loud; besides which they would not let the Tonga people be present; and to prevent them from peeping through the crevices in the reed fencing of their house, they plaistered them up with all kinds of filth, knowing that the cleanliness of the Tonga people would not then allow them even to approach. If these people are doing no harm, said the chiefs among themselves, why do they not permit us to be present? we conceal none of our ceremonies from them; why do they not expose theirs to us? Meanwhile, Morgan said to the chiefs; You see the effect of their incantations; several of you are dying every day; by and by you will be all cut off, and the king of England will take possession of your islands; for though you have the remedy in your own power you will not employ it. The chiefs took the alarm in time; they rushed upon the white men, and killed all but three, who were under the protection of Veachi, a chief of great influence.

Such was the cause of the fate of the missionaries as related by the king and confirmed to Mariner by other chiefs. On subsequent enquiry he learned that the three survivors were afterwards killed during a civil war. They might have made their escape with some natives, who invited them into a canoe going to another island, but they chose to remain; alleging that they had not quarrelled with any of the Tonga people and consequently should not be hurt. The others, however, informed them that it was the Tonga custom not only to kill an enemy but likewise all his friends and relations if possible. The missionaries replied, that as they had done no harm and meant no harm, their God would protect them. At this moment a party of natives rushed out and dispatched them with their spears.

A horrible ceremony prevailing in these islands is that of strangling children as sacrifices to the gods for the recovery of a sick relative, or a propitiation for offences committed against them. Mr. Mariner relates the circumstances of an immolation of this kind in such a manner, that we may presume him to have been an eye-witness of it. A warrior had in the heat of pursuit killed an enemy just as he was entering a consecrated enclosure, where it is the height of sacrilege to attack a foe or to spill blood. Finow, being apprized of the accident, ordered cava to be taken to the priest of his own tutelar god, for the purpose of consulting the divinity, as to what atonement was proper to be made for so heinous an offence. The priest being inspired made answer, that it was necessary to strangle a child in order to appease the anger of the gods. The chiefs then held consultations and determined that a child of one of their number by a female attendant should be the victim. On such occasions the child of a male chief is always selected as being worthier than others, and a child by an inferior female attendant, because it is not a chief; such children only being chiefs whose mothers are chiefs. In the present instance, the father gave his consent; the child was accordingly sought for, but the mother, from an anxious foreboding, had concealed it. Being nevertheless discovered by one of the men sent in search of it, he lifted it up in his arms, smiling with delight at being taken notice of. The poor mother wanted to follow, but was held back by those about her; on hearing her vioce the child began to cry, but when it arrived at the place of execution, it was delighted with the band of gnatoo that was put round its neck, and looking up in the face of the man who was about to destroy it, displayed in its beautiful countenance a smile of ineffable pleasure. This sight inspired every breast with pity, but the fear of the gods was a sentiment paramount to every other, and the executioner, as he put on the fatal bandage, could not help exclaiming: "Poor little innocent!" Two men tightened the cord by pulling at each end, and the guiltless and unsuspecting victim was soon relieved from its painful struggles. The body was then placed on a sort of hand-barrow, supported upon the shoulders of four men and carried in a procession of priests, chiefs and mataboolеs, clothed in mats, with wreaths of green leaves round their necks. In this manner it was conveyed to various houses consecrated to different gods, before each of which it was placed on the ground, all the company sitting behind it except one priest, who sat beside it and prayed aloud to the god that he would be pleased to accept this sacrifice as an atonement for the heinous sacrilege committed, that punishment might be withheld from the people. The body was then given up to the relations to be buried in the usual manner.

When Finow was at the point of death, an infant child of his by a female attendant was taken away by force while sleeping in its mother's lap, and strangled in the same manner, that the anger of the gods might be appeased and the health of its father restored.

During the residence of the missionaries, carried out by the Duff, on the island of Tonga, Tibo Moomooe, who was then king, and had attained a very advanced age, was taken dangerously ill. All possible means had been employed for his recovery, but in vain. Toogoo Ahow, his eldest son, caused a younger brother to be strangled, that his father might be restored to health. The patient nevertheless got no better; and Toogoo Ahow sent for another brother, named Colealloo, who resided at some distance from the residence of his father, upon pretext that the latter desired that he would cut off his little finger in order to propitiate the favour of the hotooas, and to obtain his recovery. Colealło accordingly repaired to his dying father. An attempt was instantly made to secure him; but he gallantly defended himself against his assailant, declaring that he would voluntarily submit to his father's will. Three strong men, however, soon overpowered and strangled him, with the assistance of his own sister. In spite of these inhuman sacrifices the old king died, and two of his wives were immolated at his *fytoca*.

It is also a very common practice to

cut off a joint of the little finger, as a sacrifice to the gods for the recovery of a superior sick relation: so that there is scarcely a person living in the Tonga islands, but has lost one or both or a considerable portion of both little fingers. Those only who can have but few superior relations, such as the near kindred of Tooitonga, Veachi, or the king, have some chance of escaping this operation, if their relatives are tolerably healthy. It does not appear to be painful; little children may be seen quarrelling for the honour of submitting to it. The finger is laid flat upon a block of wood; a knife, axe, or large stone is placed with the edge upon it, and a powerful blow with a mallet or a stone effects the separation. The wound seldom bleeds much; and to stanch it, the stump is held in the smoke arising from the combustion of fresh plucked grass. It is not washed for two days: being afterwards kept clean, it heals in two or three weeks, without any kind of application. One joint is generally taken off, but some sacrifice a smaller portion, to admit of the operation being repeated several times on the same finger, if a person has many superior relations.

During Mr. Mariner's residence in these islands, Finow, the king who had adopted him for his son, died. From the circumstantial account of the ceremonies which took place on occasion of his funeral, we extract what follows :—

The morning after the king's decease, all the chiefs and mataboolеs assembled on the *marly* of Neafoo, in the house of which the body was laid on bales of gnatoo. They were all habited in mats, and sat down waiting for the remains of the king to be brought forth. The mourners, who are always women, consisting of the female relatives, widows, mistresses, and servants of the deceased, and such other females of some rank, who choose, out of respect to officiate on the occasion, were assembled in the house and seated round the corpse. They were habited in large, old ragged mats; the more ragged the fitter for the occasion, as being more emblematical of a spirit broken down, or, as it were, torn to pieces by grief. Their eyes were swollen with weeping; the upper part of the cheeks quite black and so swollen with the frequent blows inflicted on themselves with their fists, that they could scarcely see; and their breasts were equally bruised from the same cause.

Of the chiefs and matabooles all who were particularly attached to the late king evinced their sorrow by a proceeding usual among them on the death of a relation or a great chief, when it is customary to cut and wound themselves with clubs, stones, knives, and sharp shells; one at a time, or two or three together, running into the middle of the circle formed by the spectators, to give these proofs of their grief and respect. Apostrophising the deceased, they accompanied their speeches with a wild and frantic agitation of the body, at the same time cutting and bruising their heads every two or three words with the knife or club which they held in their hands. Some, still more violent, cut their heads to the skull with such strong and frequent blows, as caused them to reel and produced a temporary loss of reason.

After these savage expressions of sorrow had continued nearly three hours, the body was laid on a bale of gnatoo, which was placed on a kind of hurdle, to be carried to the burying place, which is called *fytoca*. This includes the grave, the mount in which it is sunk, and a sort of shed erected over it. The grave of a chief's family is a vault, lined at the bottom with one large stone, having one at each side, and one at the foot and head; it is in general about eight feet long, six wide and eight deep, and is covered at top with one large stone.

The wives of the deceased and female attendants proceeded first in silent sorrow, followed by the body of Finow, then that of a favourite daughter, who had died a few days before him; next came the matabooles, and lastly the young prince and his retinue. In two hours the procession reached the place of interment, where the body was laid in a house on the *marly*, at some distance from the grave, till another and smaller house could be brought close to it. This was done in the course of an hour: the corner-posts being taken up, the four pieces which compose the building—a kind of shed of a pyramidal form, the eaves reaching within four feet of the ground—were carried by a sufficient number of men and put together at the place were it was wanted. The re-erection being finished, the body was carried into the building on the same hurdle, then removed from the latter, and laid on the bale of gnatoo: and the house was hung round with gnatoo of a deep brown colour, with black stripes, reaching from the eaves to the ground. This stuff is not chosen on account of its colour, but because it is coarse and common, emblematical of poverty and sadness: they have another sort of the same colour and pattern, but of very superior quality, which is used on occasions of rejoicing.

The women, who were now all assembled and seated round the body, began a most dismal lamentation, while a number of people, under the superintendance of a mataboole, whose office it is, were digging the place of interment at the fytoca. Having dug about ten feet, they came to a large stone vault of the kind already described. A rope was then fastened double round one end of the stone at top, for which purpose it is prevented from shutting close by means of certain bodies placed underneath, and by the main strength of one hundred and fifty or two hundred men, pulling at the two ends of the rope, it was brought up on end. The body, having been oiled with sandal-wood oil, and then wrapped in Hamoa mats, was handed down on a large bale of gnatoo into the grave: the bale was then, as is customary, taken to the above-mentioned mataboole as his perquisite. The corpse of his daughter, in the model of a canoe, was let down in like manner and placed by his side in the grave, which was capacious enough to hold thirty bodies. The great stone was then lowered down with a loud shout; on which certain mataboole and warriors ran like frantic persons round and round the fytoca, exclaiming:—"Alas! how great is our loss! Finow, thou art gone! witness this proof of our love and our loyalty!"—at the same time cutting and bruising their head with knives, clubs, and axes.

The whole company now formed into a single line, the women first and the men afterwards, but without any particular order as to rank, and proceeded towards Lico, or the back of the island, as they term it, for the purpose of fetching sand. They sang loudly the whole way, as a signal to all who might be in the road or adjacent fields to hide themselves as quickly as possible; for it is sacrilegious to be seen abroad by the procession during this part of the ceremony, and any person who might unfortunately make his appearance would presently be dispatched. Even in case of the burial of a common man, if the king himself were upon the road or in the neighbourhood of the procession while going for sand to the back of the island, he would immediately hide himself; not that his life would be in danger, but it wolud be thought sacrilegious and unlucky, the gods being supposed to be present at the time.

On arriving at the back of the island, each individual made a small basket of the leaves of the cocoa-nut tree, holding about two quarts, and filled it with sand. This done, each of the men carried two of these baskets upon a stick across the shoulder, one at each end, while the women carried but one pressed against the left hip: they then returned, with the same ceremony and by the same way they had come, to the grave. By the time they reached it, the earth recently dug out was nearly all thrown back upon the vault: the remaining small space being left to be filled up by the sand, which is always more than sufficient for the purpose, that the mount of which the fytoca consists may be strewed in like manner, such a covering being considered a great embellishment to a grave. It is moreover the custom, though no one can explain the reason why—but this is the case with many other customs of the Tonga islanders.

The sand being thus deposited, the temporary house is taken to pieces and thrown behind the fytoca into the hole out of which the earth was originally dug, to raise the mount containing the grave. Into this hole are also thrown the baskets in which the sand was brought, as well as the earth not used in filling up the grave. The ground within the house over the grave is then covered with mats, like those commonly used in dwellings, and made of the leaves of the cocoa-nut tree.

During all this time the company were seated on the green before the fytoca, still clothed in mats, and having round their necks strings of the leaves of the *ifi* tree, in token of sorrow. They now rose and went to their respective habitations, where they shaved their heads, and burned their cheeks with a small lighted roll of *tapa*, by applying it once on each cheek-bone. The place was then rubbed with an astringent berry, which causes it to bleed, and the blood they smear about the wound in a circle to the diameter of about two inches: or, those whose love for the deceased is very great, or wish it to be thought so, instead of burning the cheeks in the manner just described, rub off the skin by beating and scrubbing their cheeks with plait made of the husk of the cocoa-nut, wound round their hands, which is a very painful operation. The friction with the berry is repeated every day to make the wound bleed afresh.

The men meanwhile neglect to oil and shave themselves during the day, but perform those operations at night. After burning their cheeks and shaving their heads, they build temporary huts for their accommodation during the mourning, which lasts twenty days.

The women who have become tabooed by touching the dead body remain constantly in the fytoca, excepting when they want food, for which they retire to one or other of the temporary houses to be fed, but they sleep in the fytoca. The provisions with which these tabooed women and mourners in general are furnished, were sent in this instance by the prince. On the fifth and tenth days the supply is larger than usual, and on the twentieth day a still more abundant quantity is furnished, by way of finishing the funeral ceremony. Torches are also sent to light up the fytoca at night.

During these twenty days, if any one passes the fytoca, he must not proceed in his ordinary careless way, but with a slow pace, his head bowed down, and his hands clasped before him. If he have a burden, he must lower it and carry it in his hands or upon his bended arms; but in this case he will, if possible, make a circuit to avoid the grave. Every day some individuals approach, and sit before the grave for two or three hours, beating their faces with their fists, or stand up and bruise their heads with clubs. Such are the uniform and essential circumstances which always take place during this part of the ceremony of the interment of chiefs.

To this we may add, that Finow's chief widow went every morning, attended by her women, to cut the grass short before the grave, with knives and sharp shells, sweeping away leaves and loose blades with brooms made of the stem of the cocoa-nut leaf. They also procured sweet-scented plants, and planted them before the grave.

Early in the morning of the twentieth day after the burial of the deceased chief, all his relations, together with those who formed his household, and the tabooed women, went to the back of the island without any particular order or ceremony, to procure a number of flat pebbles, principally white but a few black, for which they made baskets on the spot, as before for the sand. Returning to the grave, they strewed the inside of the house with the white ones, as also the outside about the fytoca; the black pebbles they strewed only upon the white ones, which covered the ground directly over the body to about the length and breadth of a man, in the form of a very eccentric ellipsis. The house over the fytoca was then closed up at both ends with a reed fencing, and at the two sides with a sort of basket-work made of the young branches of the cocoa-nut tree, split and curiously interwoven, which are not taken down till the next funeral.

After the company had been feasted with provisions sent by the different chiefs, dances succeeded. During the intervals of the dances, several matabooles, warriors and others ran before the grave, cutting their heads with clubs and axes, as proofs of their fidelity to the late chief: among others, two boys of twelve or fourteen years made themselves very conspicuous by this kind of self-infliction. A still more extraordinary spectacle was exhibited by the fishermen of the late king, who advanced with three arrows stuck through each cheek in a slanting direction, so that while the points came quite through the cheek into the mouth, the other ends went over their shoulders and were kept in that situation by another arrow, the point of which was tied to the ends of the arrows passing over one shoulder, and the other end to those of the arrows passing over the other shoulder. In this horrible equipment they walked round the grave, beating their faces and heads with paddles, or pinching up the skin of the breast, and sticking a spear quite through—and all this to prove their affection for the deceased chief.

The day's ceremony, which lasted about six hours, concluded with a grand wrestling match, after which each retired to his respective home or occupation.

Some of the fytocas are remarkable specimens of the architecture of these islanders. The largest in the Friendly Islands, belonging to the family of Futtafaihe, for example, consists of a beautiful pyramidal monument, constructed entirely of hewn stones of vast size. At the base it is 156 feet in length and 140 in breadth, and it rises gradually diminishing in four steps or gradations. Though much inferior in height to the great morai in Otaheite, it surpasses the latter in the prodigious masses of stone of which it is composed. How these people with their rude implements, and destitute of mechanical resources, could have moved and wrought blocks of coral twenty-four feet long, twelve broad and two thick, is a question which we shall not pretend to answer.

At the death of Tooitonga, the great divine chief of these islands, there is such constant feasting for nearly a month, as to threaten a scarcity of certain kinds of provisions. To prevent this evil, a taboo is laid on hogs, fowls and cocoa-nuts, so that none but great chiefs may use them for food on pain of death. This taboo lasts eight months, and is then removed by a ceremony called *fuccalahi*, which is thus described by Mr. Mariner:—

The places appropriated for this ceremony were two *marlys*, Tooitonga's and Finow's, and the grave of Tooitonga. The *marly* of the latter is near his own residence where four columns of yams were erected in the following manner. Four poles, about eighteen feet long, were fixed upright in the ground about four feet distant from each other in a quadrangular form; the spaces between them all the way to the top being crossed by smaller poles about six inches asunder, lashed on by the bark of the *fow*, a species of *hibiscus*. The interior was filled with yams as the workmen proceeded; and other upright poles were afterwards lashed to the top with cross-pieces in like manner, till the column of yams was fifty or sixty feet high. On the summit was placed a cold baked pig. Four such columns were erected, one at each corner of the *marly*, the day before the ceremony, and three or four hundred hogs were killed and about half baked. The following day the hogs were carried to the king's *marly*, about a quarter of a mile off, and placed upon the ground before the house, as well as four or five wooden cars or sledges full of yams, each holding about five hundred. The people meanwhile assembled from all quarters, and sat down round the king's *marly*. Some of them occasionally rose to amuse themselves and the rest of the company by wrestling. The king and his chiefs, all dressed in cloth of the paper-mulberry-tree, were seated in the house viewing what was going forward.

Every thing being ready, the king gave notice that the ceremony was to begin. The young chiefs and warriors and those who prided themselves on their strength, then rose singly and endeavoured by turns to carry off the largest hog. The task is the more difficult on account of the greasiness of the hog, as well as its weight; but it affords much diversion to see a man embracing a large fat baked hog and endeavouring to raise it on his shoulder. Being found too heavy for one man's strength, it was carried away by two, while a third followed with its liver, and both were deposited on the ground near Tooitonga's *marly*. A similar trial was made with the second hog, but with no better success, and it was carried away in like manner by two. Those next in size afforded more sport, as being nearer to an equality with the strength of one individual. Sometimes one of these competitors had nearly raised his greasy burden on his shoulder, when it slipped through his arms, and in his endeavours to save it brought him down too. It is deemed an honour to attempt this feat, which is sometimes done by the king himself. The small hogs afforded no diversion, as they were easily lifted and carried away by one man and deposited not at the outside of Tooitonga's *marly*, with the larger hogs, but within it, whither the cars of yams were also dragged one at a time.

When every thing was thus cleared away from the king's *marly*, the spectators proceeded to the other, where they again seated themselves, whilst Tooitonga (the son of the deceased chief) presided, and the chiefs sat, out of respect, on the outside of the ring among the great body of the people. The large hogs were then brought in each by one man; for which purpose two others were allowed to lift the load and place it upon his shoulders. In this manner all the hogs and their livers were carried in and deposited in two or three rows before Tooitonga. Their number and that of the cars and piles of yams being announced to the latte., about twenty of the largest hogs were carried in like manner to the burial-place of Tooitonga, about a hundred yards distant, and deposited near the grave. One car of yams was also taken and left there.

The remainder of the provisions was then shared out. One column of yams was allotted to the king, who always distributes it among his chiefs and fighting-men; another was assigned to Veachi; the third portion, set apart for the gods, is always taken care of by the priests: and the fourth was claimed by Tooitonga. By custom and silent consent the latter appropriates the cars of yams to his own use and that of his numerous household. The hogs were disposed of in like manner, the greatest chiefs receiving a share proportionate to their rank, which they divide again among their dependents, till every man in the island obtains at least a mouthful of pork and yam. The ceremony concludes with dancing and wrestling; the assembly breaks up, and each person having secured his portion, returns home to share it with his family. From this moment the prohibition on hogs, fowls and cocoa-nuts is removed.

The hogs and yams deposited at the grave, having remained there several days till the pork was spoiled, were then shared out by Tooitonga's order to all who chose to apply for a portion. They belong by right to the principal chiefs, who, being accustomed to meat in a better state of preservation, forego their claims and allow the lower class to eat it for them. Mr. Mariner could learn no other reason for this practice than that such was the ancient custom.

It has already been mentioned that the chief called Tooitonga is considered as being of divine origin. In this character it is customary to present to him a portion of the fruits of the earth and other eatables, chiefly yams, as an offering made to the gods in his person to secure their favour and to obtain propitious crops. To this offering called *inachi* every plantation is obliged to contribute. The presentation takes place with great ceremony, the yams and other articles being sent from the other islands to that where Tooitonga chooses to reside. The day is a universal festival. The yams are carried in procession, each ornamented with red ribbons and slung upon a pole borne upon the shoulders of two men, who deposit them near the grave of the last Tooitonga, before which the king, chiefs and matabooles are seated in a semicircle. Here a mataboole addresses the gods, thanks them for their bounty and prays for a continuation of their favour. The bearers resume their burdens and carry them back in the same way they came, attended with singing and the blowing of conchs to the *murly*, where they are loosened. The other articles composing the tribute, such as dried fish, mats and guatoo, are then brought forward. Abou one-third of the whole is allotted to the gods, and removed by the servants of the priests; one half is assigned to the king, and the remainder to Tooitonga, whose portions are also carried away to be divided among their dependents.

A regular cava-party succeeds; and provisions are shared out to each chief according to his rank. The day concludes with wrestling and boxing-matches, followed by night dances, after which the people retire to their homes perfectly assured of the protection of their gods.

The quantity of provisions dealt out at this ceremony is incredible. Sometimes it happens that several great feasts are given nearly about the same time as that of the *inachi*; for instance, on occasion of the marriage or death of some great chief, or the arrival of one after a long absence from some distant island. These feasts threaten the country with scarcity, to obviate which a *taboo* or prohibition is laid on certain kinds of food, which must not be eaten for a certain length of time. A famine or war may also occasion a necessity for the imposition of this *taboo*, the ceremony attending the removal of which has been already described.

It should however be observed that the ceremony of presenting the *inachi* or first fruits of the earth was abolished, together with the title of Tooitonga, by the present king, during Mr. Mariner's residence in Vavaoo, for the purpose of cutting off all intercourse between that and the other islands.

Dancing and singing always form a part of the extraordinary festivities and ceremonies of these people; and both are very frequently practised in the evening by torch-light. These dances are generally kept up for about four hours after dark. A dance of the women is represented in the frontispiece to this volume. They are described as dancing with infinite grace, singing at the same time, while a man standing at a little distance, regulates their movements and animates them by his voice and gestures.

The principal dances of the people of Tonga have been so circumstantially described in Captain Cook's Voyages, and an account of them would occupy so much of the space to which we are confined, that we shall only advert to two not noticed by that celebrated navigator.

The first of these called *hea*, one of the most ancient dances of Tonga, is practised only by chiefs and superior matabooles, and is very difficult of execution, not only on account of the accompanying gesture but also of the singing. The chorus is composed of ten or twelve of the chiefs or principal matabooles, in the middle of whom sits one beating time on a loose flat piece of hard wood, about three feet long and an inch and a half square, fastened at one end only on a similar piece: this is struck by two small sticks, one in each hand, and produces a rattling sound. The difficulty of keeping time is owing to the extreme velocity with which they beat, especially towards the conclusion. The dancers meanwhile perform their evolutions round the chorus, exhibiting a great variety of very graceful movements with the arms and head, accompanied with expressions of countenance suitable to the character of the dance, which is consistent with the mind and habits of a superior person, and it is therefore deemed essential that every chief and mataboole should learn it. In fact it would argue great ignorance in the opinion of the Tonga people to be unaccomplished in the graceful, manly and expressive movements of this dance.

The second is a night dance called *oola*, and was formerly common among the lower classes in the Tonga islands; but some of the chiefs, during a visit to Hamoa, being pleased with the superior gracefulness with which the *oola* was danced there, have brought it into fashion among the higher ranks at Tonga. The figures are much the same as in the other dances, but the arrangement of the movements is different, and the whole dance is considerably quicker throughout.

Besides the musical instruments already mentioned, they have a sort of flute called *fango-fango*, blown by the right nostril, the left being closed by the thumb of the left hand. This instrument has generally five holes for the fingers and one underneath for the thumb. The sound is soft and grave; it is used only as an accompaniment to one species of song called *oobé*.

Most of their songs are descriptive of scenery, but some of them relate to past events or to places which are out of their reach, such as Bolotoo and Papalangi, or England. The accounts they give of the latter are very ludicrous. The poet describes among other things the animals belonging to the country, stating, that in the fields there are large pigs with horns which eat grass; and that the people there have houses which are pulled along by immense birds. The women are said to be so muffled up with dress, that a native of Tonga, entering a house, takes a lady for a bundle of *gnatoo*, which he throws on his shoulder to carry it away, when to his extreme amazement the bundle leaps down and runs off. One of these songs describes the principal circumstances attending Captain Cook's visit, and bating a little exaggeration, with tolerable accuracy, and another relates to that of the French admiral d'Entrecasteaux.

To this class of composition belongs the subjoined song, which is very frequently sung by them, or, to speak more correctly, given in a sort of recitative by either sex: it has neither rhyme, nor regular measure, although some of their songs have both. It relates to a romantic spot at the back of the island of Vavaoo, where is a wood of *toa* trees which is a favourite resort of the natives, who, after bathing in the sea, rinse themselves with the fresh water found in the hollows between the junction of the large branches and the trunks of the trees. Here also they plait flowers, which they have gathered at Matawto, a mile farther along the beach, which the women put round their necks or take home to present to their lovers, to friends, or to superior chiefs.

SONG.

While we were talking of Vavaoo tooa Licoo, the women said to us, let us repair to the back of the island to contemplate the setting sun; there let us listen to the warbling of the birds and the cooing of the wood-pigeon. We will gather flowers from the burying-place at Matawto and partake of refreshments prepared for us at Lico Onë; we will then bathe in the sea and rinse ourselves in the Vaoo Aca; we will anoint our skin in the sun with sweet-scented oil, and will plait in wreaths the flowers gathered at Matawto. And now as we stand motionless on the eminence over Anoo Manoo, the whistling of the wind among the branches of the lofty toa shall fill us with a pleasing melancholy; or our minds shall be seized with astonishment as we behold the roaring surf below, endeavouring but in vain to tear away the massive rocks. O how much happier shall we be thus employed than when engaged in the troublesome and insipid affairs of life!

Now as night comes on, we must return to the *mooa*:—but hark! hear you not the sound of the mats?—they are practising a *bo-oola* [a kind of dance by torch-light] to be performed to night on the *marly* at Tanea; let us go thither. How will the scene of rejoicing call to our minds the many festivals held there before Vavaoo was torn to pieces by war! Alas! how destructive is war! Behold! how it has covered the land with weeds, and opened untimely graves for departed heroes! Our chiefs can now no longer enjoy the sweet pleasure of wandering alone by moon-light in search of their mistresses;—but let us banish sorrow from our hearts; since we are at war we must think and act like the natives of Feejee, who first taught us this destructive art. Let us therefore enjoy the present time, for to-morrow perhaps or the next day we may die. We will dress ourselves with *shi coola* and put bands of white *tapa* round our waists; we will plait thick wreaths of *jiale* for our heads, and prepare strings of *hooni* for our necks, that their whiteness may set off the colour of our skin. Mark how the uncultivated spectators are profuse of their applause! But now the dance is over; let us remain here to night and feast and be cheerful, and to-morrow we will depart for the *mooa*. How troublesome are the young men begging for wreaths of flowers, while they say in their flattery · "See how charming these young girls look coming from Licoo! How beautiful is their skin diffusing around a fragrance like the flowery precipice of Mataloco! Let us also visit Licoo!—we will depart to-morrow."

Mr. Mariner to whom we are indebted for this specimen of the productions of the Tonga Muse, informs us that the poets of Vavaoo frequently retire for several days to the most romantic and secluded spots of the island, to indulge their genius, and return to the *mooa* with new compositions which they introduce on the first suitable opportunity.

The games and familiar pastimes in use among the people of Tonga are numerous. The most important, called *liagi*, is confined to chiefs and matabooles, who are expected to be well acquainted with it. It is played by two or four persons. In the former case they sit opposite to one another and both at once make signs with the hands. He whose turn it is to count makes one of three signs; by a sudden jerk of his arm presenting either his open hand, his closed hand, or the extended fore-finger, the others and the thumb being clenched. At the same moment also his opponent makes a sign, and if it happens to be the same, it becomes the turn of the latter to play and the first gains nothing. But if the first succeeds in making one of these three signs without his opponent making the same five successive times, he throws down one of five small sticks which he holds in his left hand. It is now the turn of the other to play, and he must endeavour to do the same; and whichever in this manner gets rid of his five sticks first wins the game. If, however, his antagonist imitates him before he can make five signs, at the fourth for example, he has a right to ask what were the three preceding movements on each side; and if his opponent cannot mention them in the order in which they occurred and give a feigned reason for every motion on both sides in the technical language of the game, according to a certain invariable system, he may begin his count again. The assigning of these supposed or artificial reasons for each move is the most difficult part of the game, because it must vary according to the order of each of the moves that preceded.

When four play, they sit as in our game of whist, but each is the antagonist of the person opposite to him; and when one has got rid of his five sticks, he assists his partner by taking one or two of his and continuing to play. The rapidity with which these motions are made is almost incredible: and the eagerness with which they play, the enthusiasm into which they work themselves, and the readiness with which the expert give the explanation of every combination of signs, excite profound astonishment in a stranger.

Catching pigeons with a net was formerly a more common sport than at present. The net used for the purpose is small with a narrow opening, affixed to the end of a rod about twelve feet long, held by the sportsman, who is concealed in a small cabin five feet high, nearly in the form of a bee-hive, having a perpendicular slit dividing it quite in half, so that he can move his rod completely from side to side. The only mode of entrance is by separating the two halves of this cabin. There are perhaps eight or nine of these receptacles, usually ranged round the upper part of a raised mount, with a sportsman in each. On the outside a trained pigeon is tied by the leg; and near at hand is stationed an attendant with another trained bird, which is tied to the end of a very long line and suffered to fly out to the extent of the string, the other end of which is held by the man. The bird describes a considerable circle in the air; its flight and the constant cooing of those below attract a number of wild pigeons, when the man by pulling his string calls in his pigeon, which perches on his finger, he then conceals himself with the other attendants in a sort of alcove at the top of the mount. The wild pigeons approach the tame ones, and the dexterous sportsmen secure them in their nets.

On these islands there is a species of bird called *kalai*, which is trained for a sort of sport, in which, however, none but the king and very great chiefs can indulge, as their training and keeping require great care and expense. The sportsman, armed with a bow and arrows, conceals himself within a large cage of wicker-work, covered with green leaves, but not so closely as to prevent his seeing the game. On the top of this cage the cock-bird is tied by the leg, and makes a noise and flaps his wings as if calling other birds to come and fight him. Within is a smaller cage containing the hen, which also makes a peculiar noise, as if answering the other on the outside. The wild birds are attracted towards the spot and are shot by the sportsman.

Each pair of birds has a man who does nothing but attend them; and without skilful management they will not make the noise requisite to attract others. These keepers are authorised to demand plantains for the birds wherever they grow, even in time of the greatest scarcity. On such occasions, if a keeper sees a fine bunch of plantains he will go and taboo it, which he does by sticking a reed in the tree and informing the proprietor that the fruit is destined for the use of the birds: and under pretence of procuring food for them these fellows frequently commit great depredations.

Shooting rats is another favourite diversion. These animals are not so large as with us, but rather between the size of a mouse and a rat: they constitute an article of food with the lower classes of the people, who, however are not allowed to shoot them, that sport being reserved for chiefs and the higher orders. Their mode of proceeding is as follows:—

A party of chiefs having resolved to take this diversion and informed their attendants what road they mean to pursue, the latter roast some cocoa-nut and walk along the appointed road, chewing the roasted nut very finely and spitting, or rather blowing, a little of it at a time out of their mouths with considerable force, yet so as not to scatter the particles much: for in that case the rat would not be tempted to stop and pick them up, and if the pieces were too large, he would run away with one piece instead of staying to eat his fill. In this manner the bait is distributed along both sides of the road, generally for about the distance of a quarter of a mile, till the men arrive at the spot where they have directions to stop. If they come to cross-roads in their way, they stick a reed in the ground in the middle of them, as a taboo or a mark of prohibition to pass that way. This mark is so effectual that if even a considerable chief were passing, on seeing the taboo, he would stop at a distance, sit down on the ground, and out of respect or politeness to his fellow chiefs, wait patiently till the shooting party had gone by: while a petty chief or one of the lower orders would risk his life by infringing this taboo. The distributors of the bait, who are called *boohi*, having arrived at the place where they are to stop, and given the orders of their chiefs to the owners of the neighbouring plantations to send a supply of pork, yams, fowls and ripe plantains, sit down to prepare cava.

The company of chiefs, having divided themselves into two parties, set out about ten minutes after the *boohi*, armed with bows and arrows, and follow one another closely in a row along the middle of the road. The two parties are mixed, the greatest chief generally going first, behind him one of the other party, and so on alternately. According to the rules of the game no person may shoot a rat that is in advance of him, excepting the man who happens to be first in the row; but any one may shoot at an animal which is abreast of or behind him. As soon as a man has discharged his arrow, whether he hits the rat or not, he changes his situation with the man behind him, so that the first may come to be last and the last first, and for the same reason two, three or more of the same party may come to be immediately behind one another. The party which first kills ten rats wins the game; and if the animals be numerous, they commonly play three or four games. The reeds placed as taboos at the cross-roads are pulled up as they pass them, that other persons coming that way may not be interrupted. After they have partaken of the refreshments provided for them, if they are disposed to pursue the sport, they send forward their attendants to prepare another piece of road.

If in the course of this game any person can obtain a good shot at a bird, he may take aim at it: if he kills the bird it counts as a rat; but whether he hits it or not, if he shoots he changes place with the one behind him. Every now and then the sportsmen stop and make a peculiar noise with the lips, like the squeaking of a rat, which frequently brings the animals out of the bushes, and they sit upright on their haunche as if listening. If a rat, alarmed by their approach, is running away, one or more cry out, *Too!* (stop)—with a sudden percussion of the tongue. This has in general the effect of making the rat stop, when he sits up and seems too much frightened to attempt his escape. When he is running away, the squeaking noise above-mentioned, instead of stopping, would cause him to run faster.

The arrows used on these occasions are nearly six feet long, which is double the length of the war-arrows: they are made of reed, headed with iron-wood, not feathered, and their great length is requisite that they may go straight enough to hit a small object, besides which it is advantageous also in taking aim through a thick bush. Each person has but two arrows, which when discharged are immediately brought back to him by one of the attendants who follow the party. The bows too are longer than those used in war, being about six feet in length, while the others are not more than four and a half: nor are they so strong as the latter, lest the difficulty of bending them should occasion a slight trembling of the hand, which would render the aim less sure.

A singular kind of sport practised by these people consists in carrying a large stone under water ten feet deep from one post to another, at the distance of seventy yards. The difficulty is to pursue a straight course; a person may thus run along the bottom twice as fast as another can swim.

A common pastime of women is to throw up five balls in such a manner as to keep four constantly in the air at once. They are discharged from the left hand, caught in the right and transferred to the left again. The players at the same time recite verses; each jaculation from the right to the left hand coinciding with the cadence of the verse; for every verse they finish without missing they count one: and sometimes seven or eight play alternately.

The character of the Hapaee people is not naturally more treacherous than that of the natives of Vavaoo or the other islands; but as they have more petty chiefs whose interest they have to consult, the opportunity for treachery is perhaps more frequent. The capture of the British privateer, Port au Prince, in which Mr. Mariner arrived at these islands in 1806, afforded proof of this quality, and had Captain Cook been acquainted with this trait in their character, he would not have misnamed them *Friendly*. They had in fact formed a deliberate plan for his destruction, and would infallibly have put it in execution but for some disagreement among themselves in regard to the best time and mode of making the attempt. Finow, grandfather of the present king, at that time tributary chief of the Hapaee islands, was not the author of this plan, but furnished advice respecting it. The other chiefs proposed to invite the captain and his officers to a grand *bomeë* or night-dance by torch-light, and at a given signal to massacre them and all their attendants. Finow objected to this, as the darkness would be unfavourable to their operations for securing the two vessels, and proposed that the attack should be made on occasion of a grand entertainment which was shortly to be given in honour of the captain's arrival; urging that after he and his party were dispatched, some of the men who were left on board would naturally come to seek him, and might be conducted to the farthest part of the island under pretext that he was there, and destroyed in like manner; and then the ships, from the weakened state of their crews, might be taken with ease.

The entertainment was accordingly

prepared, and the captain and several of his officers being invited, were present. It so happened, however, that very shortly before the concerted signal was to be given, most of the chiefs still expressed their opinion that the night-time would be more favourable than day; on which Finow, vexed at their opposition, forbade the attempt altogether. The amusements accordingly proceeded without interruption, and Captain Cook and his officers were highly delighted with their entertainment, acknowledging it to surpass every other they had received at the Friendly Islands. On so slender a thread does the life even of the most circumspect sometimes depend!

These particulars were related to Mr. Mariner at different times by several chiefs who were present, and in particular by Finow himself, the father of the present king, and son of the chief who was engaged in the conspiracy.

SANDWICH ISLANDS.

This important groupe, which has of late years strongly attracted the attention of the civilized world, consists of eleven islands lying between 18 and 22 degrees north latitude, and 150 and 160 degrees west longitude. They are called by the natives Owhyhee, Mowee, Ranai, Morotoi, Tahoorowa, Woahoo, Atooi, Neeheeheow, Oreehoua, Morotinne, and Tahoora. The last two are uninhabited. The population of the others is estimated at 400,000 souls, of which number Owhyhee alone is supposed to contain 150,000, being eighty-five miles in length, upwards of seventy broad, and nearly three hundred in circumference.

The inhabitants are undoubtedly of the same race with those of the islands south of the equator; but in their persons, language, customs and manners approach nearer to the New Zealanders, than to their less distant neighbours either of the Friendly, Society, or Marquesas Islands. They are in general above the middle size and well made, though, upon the whole, inferior in strength and activity to the Friendly Islanders.

The hair of the Sandwich Islanders would be beautiful if they allowed it to grow; it is naturally shining and black

Vol. I. p. 185.

SANDWICH ISLANDER.

Pub. by R. Ackermann London. 1824.

as jet. In general they shave the sides of the head, and allow a tuft of hair to grow at the top which falls down the back, resembling the horse-hair descending from the helmets of some of our dragoons. Some retain all their hair, and allow it to float over their shoulders or tie it very gracefully. None of them have woolly hair, but in some it is curly. Their eyes are lively and expressive, their noses mostly rather flat, though many are somewhat aquiline, as in the annexed engraving. Their mouths and lips are of middling size, and their teeth very fine; but superstition imposes on them the duty of extracting one or two in the front at the death of a friend or benefactor.

All of them are more or less tattowed. The figures in general represent birds, fans, chequers, and circles; but frequently numerous rows of goats, and these are almost always on the inner part of the arm, the leg and the thigh. Several are tattowed on one side only, which produces a very singular effect, and makes them look like men half burnt or daubed with ink from head to foot. Frequently too, from some unaccountable whim, they leave a design unfinished, as if they had changed their minds in the midst of the operation.

The same superstitious notions which deprive them of their teeth also impose on them the duty of burning their bodies, as a sign of mourning, with a red-hot iron of circular shape, in such a manner, that at first sight these dark and livid spots would be taken for so many scars. Some are covered all over with these marks.

The women are by no means so handsome as the men. They are rather short than tall; have good eyes and teeth, remarkably small and delicate feet and hands, and a sweetness and sensibility of look that render them very engaging. Their hands and arms are very neatly tattowed, and they are said to have the same operation performed on the tip of the tongue.

Children at their birth are quite black; and the most delicate females, who expose themselves least to the sun and air, display the same colour: while those who are obliged to be a good deal abroad have nearly an orange hue.

The principal part of the dress of the women is a simple garment called *pow*. It consists of a piece of cloth about one yard wide and three long, wrapped several times round the waist, with the end tucked in below, and reaching to the calf of the leg: in cold weather another piece of cloth, like a plaid, is thrown over the shoulders. Round the neck they frequently wear wreaths of the leaves of a fragrant plant, called *miri*, somewhat resembling those of the vine; also tresses of human hair to which is suspended before a piece of bone highly polished, the lower part forming a curve: these ornaments, called *palava*, are commonly made of whales' teeth, which the Americans sell at a high price to the natives. The hair necklace is plaited in small cords, which are so numerous as completely to fill the hole through which they are passed, though in general large enough to admit a man's thumb. The women comb back the hair in front, where they plaster it with a kind of lime made of burnt shells, which bleaches that round the forehead nearly white, so as to produce a strong contrast with the dark colour of the skin, as may be seen in the annexed engraving. The Russians saw many females whose hair was stained a rose colour, but could not learn how it was communicated.

Their heads are adorned with wreaths of flowers, picked from the stalk and strung on the stem of a small creeping plant. They prefer purple, yellow and white, and arrange them alternately three or four inches of one colour. This wreath, twined several times round the head, has a very elegant appearance. The women, upon the whole, take great pains in adorning their persons, for which purpose each of them is provided with a small mirror; and all ranks pay the utmost attention to personal cleanliness.

WOMAN of the SANDWICH ISLANDS.

Pub. by R. Ackermann, London, 1824.

The usual dress of the men consists merely of a small girdle, called *maro*: but upon extraordinary occasions the chiefs wear elegant cloaks and helmets of red and yellow feathers. Some of the latter reach down to their heels, and have a most magnificent appearance. They are made for the most part of red and yellow feathers, which are tied upon fine net-work. The caps are composed of the same kind of feathers, which are sometimes intermixed with black; they are secured in a kind of basket work made in the form of a helmet. Both caps and cloaks are made of various patterns and sizes. The cloaks are not all composed of the same kind of feathers, but are sometimes varied with the long tail feathers of the cock, with a border of yellow or red, and sometimes with those of the tropic bird. These articles, however, are to be seen in the possession of the principal people only.

They have also a kind of fly-flap, made of a bunch of feathers, fixed to the end of a thin piece of small and polished wood. The tail feathers of the cock are in general used for this purpose; but the better sort of people have them made of the feathers of the tropic-bird, or those of a black and yellow bird called *moho.* The handle is very frequently made of a leg or arm bone of a man whom they have killed in battle, curiously inlaid with tortoise-shell. This ornament is common to superiors of both sexes, and is highly valued.

The cloth of which their scanty garments are made, is called *tapa*. It is manufactured by the women from the bark of the paper-mulberry tree, which is first steeped in water and then beaten out with a piece of wood, grooved and furrowed like a crimping-machine. The bark is laid upon another piece of wood grooved like the former. As these two instruments are at right angles during the operation, the marks in the cloth are crossed like warp and woof.

This cloth is coloured brown, green, blue or black, with the juice of berries, laid on with a piece of turtle-shell shaped like a knife, or with a brush formed by chewing the end of a slip of bamboo: to produce a yellow, the cloth is dipped in a dye prepared by boiling the cone of a tree in water. They often paint a variety of patterns, in which they display great taste and fancy.

The *tapa*, from its texture, is liable to injury from wet, and in that state tears like moist paper: great care is therefore taken to keep it dry. Hence, when these people swim off to ships, they hold their clothes above the water with one hand, occasionally changing as it becomes fatigued.

The mats with which the floors of

the houses are covered, are also made by the women of rushes or a kind of broad-leaved grass, split at the stem, and worked into a variety of patterns.

The houses are of the simplest form: they are oblong, with very low side walls, and high, thatched roofs. They are not divided into apartments, nor do they contain any tables or seats. It is only by their size that the habitations of the chiefs are distinguished from those of the lower classes; for the same barn-like shape is universal. They are, however, kept extremely clean, and their household utensils, consisting of wooden dishes and calebashes, are hung, neatly arranged, upon the walls. While the floors of the meaner houses are bare, excepting the place for sleeping, where a few mats are spread; those of the higher classes are entirely covered with mats, many of which are worked with great elegance into different patterns. A platform at one end, raised about a yard from the ground, and extending the whole breadth of the apartment, is spread with a layer of rushes and covered with mats. This forms the sleeping-place for the family itself; but the attendants lie at the opposite end.

The two sexes never eat together: the chiefs, therefore, have always a separate-eating house, and even the lower class have one to every six or seven families for the men: the women taking their food in the same houses in which they sleep.

Few of the houses, excepting the largest, have windows; the light being admitted at the door, which is rarely closed. The dwellings of people of the higher rank are generally surrounded with paling.

Instead of candles these islanders use the *tootooee*-nut, which, from its oily nature, yields a considerable light. It grows upon a small tree, and is about the size of the horse-chesnut. After they have been baked under ground, and their shells broken off, they are put away till wanted. When employed to give light, twenty or thirty of these nuts are strung upon a slit of bamboo: each will burn five or six minutes, but they require constant trimming, and whenever a nut is nearly consumed, it is necessary to reverse the torch, that the next may catch fire. It is therefore one person's business to hold and keep it in order.

The oil obtained from this nut by pressure is well adapted for mixing with paint. The black colour, with which they paint their canoes, is produced by burning the nuts after they are pressed, and by the cinders of the torches which are carefully preserved for the purpose, reduced to powder and mixed with oil.

The women are subject to many restrictions. They are not allowed to attend the *morai* or temple, on taboo days, nor at such times to go out in a canoe. They are never permitted to eat with the men unless when at sea, and then not out of the same dish. Delicacies, such as pork, turtle, shark, cocoa-nuts, bananas or plantains are also forbidden. Dog's flesh and fish used to be the only kinds of food which they might lawfully eat; but since the introduction into the islands of sheep and goats which are not tabooed, the females have less reason to complain of their diet. Notwithstanding the rigour of these prohibitions, the women very seldom scruple to infringe them, when it can be done in secret. They frequently swim off to ships at night during the taboo, and indulge their appetites with the forbidden delicacies. Campbell relates that he once saw the queen transgressing in this particular, and that he was strictly enjoined to secrecy, as she declared that it was as much as her life was worth, should the circumstance become known. The extreme severity exercised in these respects is confirmed by the statement of Kotzebue, who, while lying in the harbour of Hanaroora, saw the body of a young female which was found floating upon the water, and learned that this poor creature, having in a state of intoxication entered the men's eating-house, was instantly strangled and her corpse thrown into the sea.

During this visit of the Russians, the wife of a chief remained on board an American vessel too long, and did not return to land till after sun-set on a taboo day. As that is the time when the taboo begins, the people waited on the shore for her and rushed upon her to execute summary punishment for the transgression. The Americans and English who accompanied her, prepared for such an attack, by their joint exertions saved her life for that time.

The men of the lower class in general content themselves with one wife, but they are not confined to that number. The chiefs frequently have several. Campbell mentions one who, in his time, had no fewer than fifteen.

The conjugal tie is extremely loose; either party having a right to quit the other when they tire or disagree. The men are very jealous of any attentions paid by natives to their wives, but on the part of foreigners such attentions are accounted the surest proof of friendship. The virtue of the king's wives is guarded by attendants of both sexes, who are answerable with their lives for the faithful execution of their trust. We are assured that a nephew of Tammeamea, the late king, was strangled by his nobles for a criminal intimacy with one of his wives.

Immediately after child-birth, women are obliged to retire to the woods, and there to remain ten days, during which they must not be seen by the men. The king's wives have a house for the purpose of retirement on such occasions; the other females in general have no shelter but what the woods afford.

The people here, like those of the Friendly Islands, intoxicate themselves with *cava* or *ava*, which is prepared in the same manner, and to the effect of which are attributed the cutaneous disorders with which the South Sea islanders are afflicted. Some of the chiefs, who are in general very corpulent, drink it to reduce their bulk: but most of them manifest a stronger predilection for rum, with which they are supplied by the Americans. Campbell states that they likewise distil themselves a liquor to which they give that name not from the sugar-cane but another indigenous plant.

The inhabitants of the Sandwich Islands are composed of four great classes, exactly corresponding with those into which the natives of the Friendly Islands are divided. The land belongs to the king, under whom it is held by the *erees*, or chiefs, as hereditary but inalienable fiefs. Distinguished chiefs are placed as governors over different islands and territories, but the king receives tribute from the whole land. The common people are the dependants of some chief, for whom they cultivate the ground or work at other employments and by whom they are supported in old age. They are not, however, slaves, or attached to the soil, but at liberty to change masters when they think proper.

The principal duties of the executive power are entrusted to the priests. It is by them that the laws are enforced and the revenues of the king collected. Superstition is the most powerful engine for effecting these purposes, actual punishments being rare. During Campbell's residence in the Sandwich Islands, he knew but one instance of capital punishment, which was inflicted on a man who had violated the sanctity of the morai. Having drunk to intoxication, he quitted the sacred edifice during taboo time and entered the house of a woman: but was immediately seized and carried back to the morai where his eyes were put out. Having been left two days in this state, he was then strangled and his body exposed before the principal idol.

The mode of detecting robbery or theft affords a strong instance of the power of superstition over their minds. The party who has sustained the loss repairs to a priest to whom he presents a pig, and states the case. The priest rubs together two pieces of green wood till a kind of powder resembling snuff is produced by the friction. This powder is so hot, that, on being placed in dry grass and blown upon, it takes fire; with this a large pile of wood is kindled, and allowed to burn a certain time. He then takes three *tootooee* nuts and having broken the shells, throws one of the kernels into the fire, at the same time saying a prayer; and while the nut is crackling in the fire, he repeats the words: "Kill the fellow!" The like ceremonies are repeated with each of the nuts, should the thief fail to appear before they are consumed. This, however, but rarely happens. The culprit generally makes his appearance with the stolen property which is restored to the owner, and the offence is punished by a fine of four pigs. He is then dismissed with a caution not to commit a similar crime in future upon pain of a more severe penalty. The pigs are taken to the morai, where they are offered as sacrifices and afterwards consumed by the priests.

Should it happen that the unfortunate criminal does not appear during the awful ceremony, his fate is inevitable ; no gift can avert the effect of the prayer or appease the anger of the *etooah*, or god. The circumstance is reported to the king and proclamation made that a certain person has been robbed and that the guilty persons have been prayed to death. So firm is their belief in the power of these prayers, that the culprit pines away, refuses sustenance and at last falls a victim to his own credulity.

Their principal god, to whom they attribute the creation of the world, is called *etooah*; and they have seven or eight subordinate deities, whose images are in the morai and to whom offerings are likewise made.

They have a tradition of a general deluge. According to their account, the sea once overflowed the whole earth excepting the mountain called Mouna Kaa in Owhyhee, and swept away all the inhabitants but one pair, who saved themselves on that mountain and were the progenitors of the present race of mankind. According to this hypothesis we Europeans are the descendants of Sandwich Islanders.

Their *morais* or places of worship consist of one large house or temple, with some smaller ones round it, containing the images of their inferior gods. The tabooed or consecrated precincts are marked by four square posts placed about thirty or forty yards from the edifice. Across one end of the inside of the principal house there is a screen or curtain of white cloth, behind which is placed the image of *Etooah*. On the outside are ranged several hideously ugly wooden idols, the mouths of which are stuck full of dog's teeth.

The *morai*, or temple, of the king,

Morai. of the King of the Sandwich Islands.

Pub. by R. Ackermann. London 1824.

represented in the annexed engraving, is surrounded by a dozen grotesque colossal wooden images, from three to eight feet in height; some standing and others thrown down, and half covered with stones and rubbish, at the time of Captain Freycinet's visit. The figure with a bird on its head is said to represent the god of war. To these divinities are offered hogs, bananas, cocoa-nuts, and even human sacrifices; but, according to the concurrent testimony of navigators, the latter are always criminals. A fence formed of the spines of the cocoa-nut tree, two feet and a half high, encompasses the morai, and in the centre there is a building, in which, according to the last French circumnavigators, a variety of new and elegant articles of furniture were carelessly kept.

Arago, who accompanied Captain Freycinet, describes another morai, about a mile from the town of Kayerooa, which is an inclosure of about a hundred square yards, surrounded by a wooden railing four feet high. The wooden idol at the entrance is very carefully carved and larger than the others. The head is of enormous size, forming nearly a third of the whole figure, which, including a post about two feet high, measures ten or twelve feet. Offerings are put into the mouth. The other idols are placed on the sides of the inclosure, very close to each other; five on the left, the last of which has its head covered with a sort of long pointed hood, painted red; and six on the right, the largest of which is extended on the ground, and its neighbour half overthrown.

The religious ceremonies take place about four times a month, and last from sun-set till sun-rise on the second succeeding morning: during this interval, which is spent in prayer, in sacrificing pigs, and feasting upon the sacrifices and in conversation, no person is allowed to pass the bounds of the morai. The most profound silence is preserved while the priest is engaged in prayer; indeed, the least noise within or near the morai at such times is considered as offensive to the deity; for which reason, whenever the king enters it, proclamation is made by the public crier, ordering all animals near it to be confined, otherwise they will be seized and offered in sacrifice. These ceremonies are attended chiefly by persons of the superior class, and women are never permitted to be present at them.

During the period of the year, called *macaheite*, which falls in our November, and lasts a whole month, the priests are engaged in collecting the taxes, which are paid by the chiefs in proportion to the extent of their domains, and consist of mats, feathers, and other native produce. The people celebrate this festival by dancing, wrestling, and other amusements. The king remains the whole time in the morai. Before he enters it a singular ceremony takes place. He is obliged to stand till three spears are darted at him. The first he must catch in his hand, and with it ward off the two others. This is not a mere formality; for the spear is thrown with the utmost force, and should the king lose his life there is no help for it. The late king was so dexterous in the use of the spear, that he ran little risk in thus exposing himself. Vancouver informs us, that in a sham fight, he saw him ward off six spears hurled at him almost at the same instant. Three he caught as they were flying with one hand; two he broke by parrying them with his spear; and the sixth, owing to a slight inclination of the body, passed harmless.

A recent voyager, on the authority of resident Europeans, informs us that the punishment of death is preceded by a fast of two hours. The criminal is then conducted bound to a morai, at the entrance of which the priest is waiting for him. After he has pronounced a certain form of words, two or three persons extend the culprit on his back upon a plank, with his head resting on a stone; while the executioner, selected for the occasion from among the most athletic of the spectators, dispatches him by striking his forehead with a club. His body is either interred immediately, or left for the birds of prey, according to the will of the priest and the nature of the crime.

Another mode is as follows: the criminal is fastened to a cocoa-nut tree with his back against it. Two men passing a cord round his neck, draw it with all their force, one at each end, behind the culprit, who is soon strangled.

Although such executions are of very rare occurrence, they are attended by very few spectators.

In the Sandwich Islands, according to Campbell, the bodies of the dead are always disposed of secretly, and he never could learn where they were interred. The queen, he tells us, preserved the bones of her father carefully wrapt up in a piece of cloth. When she slept in her own house they were placed by her side: and in her absence they were laid on a feather-bed which she had received from the captain of a ship, and which was used for that purpose only. When Campbell asked the reason of this custom, she replied: "It was because she loved her father so dearly." Mariner also saw these bones, and on enquiry found that this was not a custom with the queen only, but a common practice among these Islanders.

Their instruments of war are spears, daggers, clubs and slings, and for defensive armour they wear strong mats which are not easily penetrated by such weapons as theirs. The daggers are made of heavy, black wood, resembling ebony; being from one to two feet long, with a string passing through the handle for the purpose of suspending the weapon from the arm. Some of these may be called double daggers, being sharp at each end and having a handle in the middle to strike different ways. It is not improbable, however, that all these weapons will soon be superseded by the use of fire-arms: for so early as the year 1810 king Tameamea had a regular guard of about fifty men who did duty about his residence. They were armed with muskets and bayonets, but had no uniform: their cartridge-boxes, made by native workmen, were of wood, rounded to the shape of the body, and covered with hide. In exercising, rapidity and not precision seemed to be their chief object.

The flat land along the coast of these islands is highly cultivated. The most common crops are taro-root, yams, and sweet potatoes; but the former is the chief object of their industry, being the principal article of food among all classes of the inhabitants. The mode of culture is extremely laborious, as it is necessary to have the whole field laid under water. It is raised in small patches seldom above one hundred feet square, surrounded by embankments, generally about six feet high, the sides of which are planted with sugar-canes, and having a walk at the top. The fields are intersected by channels for the supply of the water necessary to cover them.

The ground is first carefully dug and levelled with a wooden spade, which the labourers use squatting on their hams and heels; and it is then trodden down with the feet till it is close enough to retain water. The mode of propagation is by means of small cuttings from the upper part of the root, with leaves adhering to them. After these are planted, the water is admitted till it covers the surface to the depth of a foot or eighteen inches. Each of the plants sends forth a number of shoots or suckers all round, and in about nine months they are ready for taking up. Notwithstanding the labour of this mode of culture, especially as those engaged in its operations are almost constantly up to the middle in mud, Campbell assures us that he has often seen the late king working hard in a taro-patch. He adds, that these islanders are, without exception, the most industrious people he ever saw. So great was the difference in this respect between them and most of the Europeans resident on the islands, that the natives were not a little puzzled to account for the diligence with which

William Davis, a Welshman, cultivated his fields: and they at length explained it by supposing that he was one of their own countrymen, who had gone to Caheite, or England, after his death and come back again to his native land.

When the islands were discovered, pigs and dogs were the only useful animals upon them; but the late king paid so much attention to the preservation of the breeds left by Vancouver and other navigators, that they are likely soon to possess an abundant stock of horned cattle, horses, sheep, and goats. Campbell states that in 1810, there was, as he was informed, many hundreds of cattle running wild in Owhyhee, and several in a domestic state. The king had introduced the breed into Woahoo, and had then a herd of nine or ten on the north side of the island. Sheep and goats were already very numerous, and several individuals had large flocks of them. The queen's consisted of one hundred and fifty head; and Manina, the king's interpreter, had several hundred. The king had five horses, of which he was very fond, and he used frequently to ride out upon them.

To the industry of a Spaniard, named Marini, the Sandwich Islands in general, and Woahoo, where he resides, in particular, are greatly indebted. He has assiduously introduced and multiplied various kinds of animals and plants. He possesses a numerous herd of horned cattle, sheep, and about twenty horses; also asses and mules, which are found still more serviceable in the mountains composing the interior of these islands. Many foreign trees and plants are reared in his plantations, and several that he has introduced are already found every where in a wild state. After many fruitless attempts he has at length succeeded in growing rice from grain brought from China. On the hill overlooking the town of Hannaroora, he has planted a vineyard of considerable extent, which produces fine grapes; but he is said to be yet unskilled in the art of making good wine.

Tobacco, which all the nations of the earth have been equally ready to adopt, is now grown in these islands, from seed originally introduced by the Americans: it is very strong. The natives have become such determined smokers, that not only are women frequently seen with pipes in their mouths, but according to Kotzebue, children begin to smoke before they can walk alone, and some adults have carried the practice to such excess, as to fall down senseless and even to die in consequence.

Among the most common amusements of these islanders are dancing, wrestling and foot-races. The dances of the men are described as being extremely graceful. They do not move the feet much, merely making one step forward, backward, or on one side: but the head, arms and body are in a constant motion, which so far from being vehement or extravagant, displays great harmony and elegance. In general not more than three men dance together before a circle of spectators; but the women sometimes assemble to the number of fifty or more, forming solid squares, ten or twelve deep; throwing themselves into a variety of attitudes, sometimes all squatting, at others all springing up at the same moment. A man in front having strings of shells on his ancles and wrists, with which he marks the time, acts as fugle-man.

Both men and women have a particular costume for dancing · that of the men is represented in the opposite plate. They have in their left hands small bucklers covered with feathers, to each of which is attached a small calebash, with a number of pebbles in the inside. These, when the buckler is shaken, produce a noise that is not disagreeable, especially when it is done in cadence with the air of the dance. The musicians who accompany the dancers, have a large empty calebash in the left hand, which they gradually raise and then drop on the ground.

DANCER of the SANDWICH ISLANDS.

Pub. by R. Ackermann, London, 1824.

With the right they beat a small drum made of the shell of a cocoa-nut, covered with shark's skin. Besides these instruments the women have pieces of wood which they strike in cadence against one another, or pieces of reed, having transverse notches, over which are passed small sticks that produce a harsh noise like the sound of a rattle. The men wear on the fore-arm ornaments of bone, and round their legs several strings of shells which rattle when they stir.

It is a favourite amusement with them to throw the top shoots of the sugar-cane at each other, and to catch them in their flight—a practice which tends to make them very expert in the use of the spear.

Another diversion of which they are very fond, is flying kites, made of *tapa*, of the usual shape, but of uncommon size, many of them being fifteen or sixteen feet high and six or seven broad: they frequently require three or four hundred fathom of line, and as they are too powerful to be held, it is found necessary to tie them to trees. Campbell remarks, that the only employment in which he ever saw queen Tamena engaged was in making these kites.

They have a game somewhat resembling draughts, but more complicated. It is played upon a board about twenty-two inches by fourteen, painted black, with white spots, on which the men are placed. These consist of black and white pebbles, eighteen on each side, and the game is won by the capture of the adversary's pieces.

Campbell gives us a curious account of an attempt made during his residence on these islands, to introduce theatrical amusements among their inhabitants. A theatre, says he, was erected under the direction of James Beattie, the king's block-maker, who had once been on the stage in England. The scenes, representing a castle and a forest, were constructed of different coloured pieces of *tapa*, cut out and pasted together. I was present on one occasion at the performance of Oscar and Malvina. This piece was originally a pantomime, but here it had words written for it by Beattie. The part of Malvina was performed by the (native) wife of Isaac Davis, a Welshman, who had resided twenty years in the Sandwich Islands. As her knowledge of the English language was very limited, extending no farther than to the words yes and no, her speeches were confined to those monosyllables. She acted her part, nevertheless, with great applause. The Fingalian heroes were represented by natives clothed in the Highland garb, also made of *tapa*, and armed with muskets. The audience did not seem to understand the play well, but were greatly delighted with the after-piece, representing a naval engagement. The ships were armed with bamboo cannon, and each of them fired a broadside by means of a train of thread dipped in salt-petre, which communicated with each gun, after which one of the vessels blew up. Unfortunately the explosion set fire to the forest and had nearly consumed the theatre.

From their earliest years these islanders spend much of their spare time in the water, and constant practice renders them so dexterous that they seem quite as much at their ease in that element as on land. They frequently swim several miles off to ships, sometimes resting upon a plank shaped like an anchor-stock, and paddling with their hands, but more commonly without any assistance whatever.

In every article of their manufacture these islanders display an extraordinary degree of neatness and ingenuity, considering the simplicity of the tools with which they work. The tool in most common use is a species of adze or tomahawk, called *toë*: it was formerly made of hard polished stone, which is now universally superseded by iron. They form it by lashing a thin iron plate from one to four inches broad and five or six long, to a branch of a tree, which has a piece of the stem attached to it. Plane-irons are in great request for this purpose; but the *toë* is frequently made of an old hoop. This tool, with a piece of coral for a file, is almost the only one used in the construction of houses, canoes, and wooden implements. With these simple tools, circular wooden bowls, containing from half a pint to five or six gallons, are as neatly made as if they had been turned in a lathe. The facility with which they learn the useful arts of their visitors is astonishing. Many of the natives are now employed as carpenters, coopers, blacksmiths, and tailors, and their work is as perfect as that of Europeans. While Campbell was among them, there were in the king's forge none but native blacksmiths, who had been taught by the armourer of a ship.

The Sandwich Islands were among the discoveries of our great navigator, Captain Cook, who named them after his noble patron, the Earl of Sandwich, who then presided at the board of Admiralty. Here too, in Owhyhee, his useful career was prematurely terminated in a manner and under circumstances with which none of our readers can we presume be unacquinted. It is not so well known that the people of these islands, though they actually took away his life have paid and still continue to pay the highest honours to his memory, esteeming him as one sent by the gods to civilize them and to whom they are indebted for the most important blessings they enjoy. They have still in their possession the greatest part of his bones which they hold sacred: they are deposited in a house consecrated to a god, and are annually carried in procession to many other consecrated houses, where the priest thanks the gods for having sent to them so great a man.

Such at least is the account given to Mr. Mariner at Woahoo by Harebottle, an Englishman, many years resident in these islands, who added by way of explanation, that the natives delivered up very few of the Captain's bones, but substituted those of some of his men who fell on that melancholy occasion. From natives themselves the same voyager learned, that they had no idea that Captain Cook could be killed, for they considered him as a supernatural being and were astonished when they saw him fall. The man who struck the fatal blow was a carpenter, who, living a considerable distance up the country, was not even acquainted with the person of the extraordinary being, whose death was deeply deplored by the king and principal chiefs. The flesh of the illustrious victim was shared out to different gods and afterwards burned, and the bones were disposed of in the manner related above.

The narrative of Captain Cook's

third voyage introduced to us a young chief, whose ambition, seconded by his politic encouragement of European settlers, had raised him at the time of Vancouver's visit in 1794 to the sovereignty of Owhyhee. With a view probably to confirm and consolidate his newly-acquired authority by securing a powerful ally, Tammeamea, in an assembly of his principal chiefs on board Vancouver's ship, the Discovery, made a formal cession of the island to the king of Great Britain, with the understanding that no interference should take place in the religion, government and domestic economy of the natives. He now began to direct his attention towards the creation of a naval force for the purpose of prosecuting his plans against the other islands which were at this time governed by independent chiefs. He purchased fire-arms and ships of the English and Americans, built smaller vessels himself, and subdued the islands of Mowee, Morotoi and Woahoo, in the latter of which he afterwards fixed his residence. The chief of Atooee voluntarily submitted, and in short the whole groupe of the Sandwich Islands was reduced under his authority.

At the period of Kotzebue's visit, a large three-masted vessel, and a handsome brig belonging to Tammeamea, who had purchased them in exchange for sandal-wood, were lying in the harbour of Hanarooroo in the island of Woahoo. The first, called the Albatross, which then served to convey provisions from Woahoo to Owhyhee, was destined to open a trade under the king's flag with China. The brig called after the name of the Queen Kahumanna was capable of carrying eighteen guns and serving as a ship of war. She was originally a French privateer, but had been taken by our cruisers and sold to English merchants by whom she was disposed of to Tammeamea. Her second officer, named Adams, entered into the king's service and became her commander, with a salary of fifty piastres a month and a gratuitous supply of all kinds of provisions. The crew was composed of six Europeans and several natives.

From the pains taken by that monarch to create a navy and from his encouragement of commerce and navigation, the Sandwich Islands now contain many good native seamen. Some of these have been as far as China, the north-west coast of America, Kamtschatka and even the United States of America. In a short time they become useful hands and so continue as long as they remain in warm climates; but they are not capable of withstanding the effects of cold.

The harbour of Hanarooroo, the residence of the late king, is defended by a fort which mounts thirty guns, eight twelve and sixteen pounders. On its walls waves the royal flag, which consists of seven stripes, alternately blue, red and white, and at the upper corner next to the flag-staff the English jack.

Kotzebue has given a whimsical account of king Tammeamea and his court. He was received by him in a hut built after the fashion of the country, consisting of one apartment, though he had habitations of stone in the European style. His dress consisted of a white shirt, blue pantaloons, a red waistcoat and a coloured neckcloth. On great occasions, however, he dressed very splendidly, having several laced uniforms, but the national dress of his subjects was worn in ordinary both by himself and his chiefs. The Russians were accommodated with European chairs, but the distinguished natives who were present at this interview were seated on the ground. Their costume appeared still more extraordinary than the king's, consisting of a black frock coat and a small white hat. These black frocks look very odd on the naked body, and besides this they seldom fit, being brought from America, where the people in general are neither so tall nor so corpulent as the chiefs of the Sandwich Islands. One of these attendants had the waist half way up his back: the coat could not have been buttoned without the greatest difficulty, he perspired copiously, and was evidently miserable in his confinement, but fashion forbade him to release himself from its trammels. The sentinels at the door were stark naked: each had a cartouch-box and a pair of pistols fastened about his body, and a gun in his hand.

The captain next visited Tammeamea's favourite wife Kahumanna, with whom he found the two other wives of the king. The house in which she resides is very neatly built and very clean within· the floor, on which the three ladies were seated in the Asiatic manner, was covered with mats and their persons were enveloped in the finest stuffs of the country. The chief employment of the royal dames consists in smoking tobacco, combing their hair, driving away the flies with a fan and eating. Their pipes have no tubes attached to them; but the heads which are constantly hanging by their sides are of the size of the largest German pipes: they are made of a dark-coloured wood and mounted with copper. Kahumanna took with great zest a few whiffs from the pipe, swallowed a part of the smoke and expelled the remainder through the nostrils: she then offered it to Captain Kotzebue who declined the honour, ,on which she handed it to her neighbour, who soon resigned it to the third. As soon as the pipe was emptied in this manner it was filled afresh and passed round as before. All three queens were very large corpulent women, upwards of fifty, and seemed never to have possessed any claims to beauty. Their dress was distinguished by several silk handkerchiefs from that of the other females.

The king's daughter, a tolerably handsome girl, was seated on a mat before the door; behind her stood a little black page who held a silk parasol over her head to screen her from the sun, while two other boys drove away the flies with bunches of red feathers: the whole forming a pleasing groupe.

Tammeamea expired in the island of Owhyhee in March 1819. Aware of the approach of death he summoned around him the chiefs of the different islands and exhorted them to hold sacred his useful institutions, "for which," said he, "we are indebted to the white men who have come hither to reside among us." He enjoined them most particularly to respect these strangers, to hold their property inviolate, and to continue to them the rights and privileges which he had conferred. He appointed his son Rio-Rio, his successor, and left about half a million of dollars, chiefly accumulated by traffic with Europeans, besides goods and armed merchant-vessels to a like amount.

Near the temple or morai in Karakakooa Bay, Owhyhee, is an edifice beneath which are interred the remains of Tammeamea, whose name is never pronounced by his people but with the greatest veneration. It is a building thirty feet square, solidly constructed of canes, the corners projecting a little. The door is of wood, four feet and a half high and fastened by a large padlock. Two stakes placed crosswise at the entrance show that the place is tabooed and that all access is consequently prohibited.

The predecessor of Tammeamea in Owhyhee was considered so sacred that he was not allowed to be seen by day he shewed himself only in the night; and if any person accidentally saw him by day-light he was immediately put to death.

At the death of the king and distinguished chiefs human victims are killed; but these are always of the lowest class. In certain families of this class the fate of dying with the different members of certain distinguished families is said to be hereditary. The victims know their destination and their lot seems to have no terrors for them. The progressive spirit of the times has, however, almost rendered this custom obsolete. At the death of queen Kahumanna's mother when three persons offered themselves to fulfil their destination, Kareimokoo, who may be styled the viceroy of Woahoo, and who is familiarly named by the English Billy Pitt, on account of his influence with the late king, would not permit the sacrifice. They still sacrifice culprits to the gods, as we sacrifice them in Europe to justice. The custom of eating human flesh had ceased long before the time of Captain Cook.

Campbell informs us that the respect paid to the king's person, his house, and even his food, formed a remarkable contrast to the simplicity of his mode of living. Whenever he passed his subjects were obliged to uncover their heads and shoulders. The same ceremony took place on their entering or even passing his residence; and every house which he entered was ever afterwards honoured with the same marks of respect. When his food was carrying from the cooking-house, every person within hearing of the call *Noho*, "Sit down!" given by the bearers, was obliged to uncover himself and squat down on his hams. This ceremony was particularly inconvenient when the water used in the king's house was carried past: there being none of a good quality near Hanarooroo, it was necesssary to bring it from the mountains, a distance of five miles. Those who carried it in calebashes were obliged to call out *Noho!* when any person appeared in sight. In order, however, not to detain his majesty's subjects in so unpleasant an attitude, they ran past as quickly as they could. White people were not required to pay these honours though they were scrupulously exacted from the natives.

From the reports of the Missionary Society it appears that, in consequence of the accounts successively brought to the Sandwich Islands of the change produced at Otaheite and the neighbouring isles, the chiefs of Owhyhee, Woahoo and Atooi renounced their idols in the year 1819, and committed them with every vestige of idolatry to the flames. Soon after this event missionaries from the United States of North America arrived at Woahoo.

From the same source we learn also that a deputation lately sent by the Missionary Society to the South Sea were induced to accept the offer of a free passage from Huaheine to the Sandwich Islands, made to them in February 1822 by Captain Kent of his majesty's cutter Mermaid, and took with them a missionary and two native teachers, with the intention of leaving them at the Marquesas on their return.

The arrival of these strangers seems to have been the occasion of the commencement of a change in this northern groupe as extraordinary as that which has taken place in the Society Islands. No sooner had they come to an anchor in Karakakooa Bay, than the governor of Owhyhee went on board the Mermaid. This chief, whose name is Kooakeenee and who is brother-in-law to king Rio-rio, is described by Arago, as being six feet three inches high, with a mild and pleasing countenance, and acquainted with the English language. Being informed that there were missionaries in the ship, he expressed an earnest desire that they might settle at Owhyhee, as he wished to be instructed in the knowledge of the true God, concerning whom he had heard nothing but from an Otaheitean, who had indeed given him much information on the subject of the christian religion. The deputation went several times on shore during the stay of the vessel at Owhyhee and conversed with the natives, who every where expressed their desire to receive similar instruction and also to be taught to read and write. They likewise visited the governor at his own house and had much conversation with him. He was very solicitous to be made acquainted with the state of things in the Society Islands, and appeared to be much gratified with the information which he received in answer to his inquiries.

The real errand of Captain Kent was to deliver to Tammeamea, the late king of the Sandwich Islands, a schooner, presented to him by his Britannic Majesty, as a token of acknowledgment for the uniform attention paid by him to English vessels touching at any of his Islands for refreshments. Before these instructions could be carried into effect, Tammeamea had been succeeded by his son, Rio-rio, who resided at Hanarooroo, in the island of Woahoo, whither the Mermaid of course proceeded.

The king or principal chief of the island of Atooi, was at this time at Woahoo. An Otaheitean, who had been absent from his native country for at least thirty years, having left it when a boy, held the post of steward to a brother of the queen's. This man, hearing that some of his countrymen had arrived at Woahoo, went and invited them to reside at his house. In conversation with them concerning his family, it was discovered that the wife of Auna, one of the newly arrived native teachers, was his own sister. In consequence of this discovery, the king and queen of Atooi invited the Otaheiteans to be their guests, and made particular enquiries concerning the state of things in the Society Islands. They had been previously told that the kings, chiefs, and natives in general of those islands were in the most abject and degraded condition; but the statements of Auna thoroughly convinced them of the falsehood of this report; and they and their attendants expressed a wish to be taught to read and write.

In a few weeks the deputation received a message from the king and queen of Atooi and many of the chiefs, earnestly soliciting that Auna and his wife might be permitted to remain in the Sandwich Islands to instruct the people "in the word of God and the good way to heaven"; and Auna added his request, that the missionary, Mr. Ellis, would in that case consent to settle in the islands. After conferring with the American missionaries and Mr. Ellis, the deputation agreed that the Otaheitean teachers should remain in the islands and that Mr. Ellis should fetch his family from Huaheine, and settle in the country. This arrangement received the approbation of Rio-

rio and his principal chiefs, and proved highly gratifying to the king and queen of Atooi. Mr. Ellis accordingly applied himself with diligence to the study of the language of the Sandwich Islands, and from its close affinity to the Otaheitean, he was able in two months to speak the former, and to preach in it with ease and fluency. From the same cause the Otaheitean teachers could soon converse familiarly with the natives of the Sandwich Islands.

Although the natives in general hesitated as to their choice between paganism and christianity, it was yet evident that the bias of their minds was in favour of the latter. Still they were reluctant to declare their preference till they had first obtained the sanction either of the example or the authority of the king. Rio-rio, on the other hand, had assured the deputation shortly after their arrival, that he would after a while "tell his people that they must all learn the *good word* and worship Jehovah, but that the missionaries must first teach *him* and make themselves acquainted with the language."

The king proved as good as his word, and the proficiency of the missionary in the language, the increasing number of his hearers, and the improvement visible in the conduct of many, seem to have accelerated his determination. In a letter of the 10th of August, 1822, the deputation announced that his decision had been made in favour of the Gospel; and that the king, his wives and a great number of chiefs were daily receiving instruction in reading and writing, so that the royal residence and the houses of the chiefs had the appearance of school-rooms.

While the deputation continued at the Sandwich Islands, the king and queen of Atooi, accompanied by Auna, made a tour round the island of Owhyhee, during which above a hundred idols were discovered at one place in caves situated among the mountains: these were all burned together; and many more were destroyed in other parts of the island during this tour. When idolatry was formally abolished in 1819, these images were concealed by those who were adverse to the change.

The deputation left Woahoo on the 22d of August, on their return to Huaheine. Mr. Ellis accompanied them for the purpose of removing his family to his new destination.

To this account we subjoin the following particulars relative to the present state of the Sandwich Islands, given on the authority of an American captain, named Gardner, who likewise visited them in 1822 :—

The Sandwich Islands begin to have a considerable traffic, and the natives are making rapid strides in civilization. For several years past they have been visited by so many English and Americans that they are gradually adopting their manners and relinquishing their own. The bow and the spear are no longer to be seen; the harsh war sound of the Triton's horn has ceased to be heard, as have also the screams of the victim destined to the slaughter. Idolatry is at an end: the bells of the churches alone break the silence of the sabbath, and the mild beams of christianity have already begun to operate on these children of nature. Several missionaries from the United States reside among them: they have founded a school where many of the youth receive instruction in reading, writing, drawing, &c. which, together with the religious exhortations at church, contribute daily to exalt and refine the moral character of these simple people.

The natives already possess ten ships built and equipped in the European fashion, none of which is under 120 tons burden, besides a great number of schooners and sloops employed in the conveyance of sandal-wood and provisions from one island to another. Most of them are manned by natives who make excellent sailors. While Captain Gardner was at Woahoo, one of their vessels manned entirely by natives, but commanded by a white man, returned from a voyage to Kamtschatka. In exchange for a cargo of salt which she had carried to the governor of that peninsula, she brought back smoked salmon, cables, linen, hardware, and other articles. The governor transmitted by this ship a written grant of a large tract of land to the king of the Sandwich Islands.

THE MARQUESAS.

This archipelago is composed of two groupes of islands, called the Old and the New Marquesas. The former were originally discovered about the conclusion of the sixteenth century, by the Spaniards, who gave to four of them the names of La Magdalena, St. Pedro, Santa Christiana, and La Dominica. The fifth received from Captain Cook, the appellation of Hood's Island.

The New Marquesas have been principally visited by Americans, by whom they have been denominated Washington's Islands. Of Nukahiwa, the largest of them, from sixty to seventy miles in circumference, we have pretty circumstantial accounts by the Russian Captain Krusenstern, and Mr. Langsdorff who accompanied him.

Captain Cook pronounces the inhabitants of these islands collectively to be the finest race of people in the South Sea. The Russian navigators coincide in his opinion. Judging, says Langsdorff, from the accounts of all who have visited the Friendly and Society Islands, I am inclined to think that the people of the Marquesas excel in beauty and stature, in regularity of features and in colour, all the other South Sea Islanders. The men are almost all tall, robust, and well made. Few were so fat and unwieldy as the Otaheiteans, none so lean and meagre as the people of Easter Island. We did not see a single cripple or deformed person, but such general beauty and regularity of figure as greatly excited our astonishment. Many of them, had they been placed beside the most celebrated master-pieces of antiquity, would not have lost by the comparison. Their beards are commonly black, shining and thin, as they are in the habit of plucking up the hairs by the roots. The hair is commonly long, curly, strong and black; but that of a few somewhat less dark.

A man named Mufau particularly attracted the notice of the Russians by his extraordinary stature, the vast strength of his body and the admirable proportion of his limbs and muscles: he was then twenty years old, six feet eight inches high, English measure, and perfectly proportioned. The chief of a neighbouring island, as they were assured, was a head taller, with equally fine proportions.

The expression of countenance in these people is generally pleasing, open and animated, and their dark eyes are full of vivacity. The women in general are much smaller than the men, but well-proportioned. Their faces are rather full and round than long; they have large sparkling eyes, clear complexions, very fine teeth, regular and expressive features and mostly black, curly hair.

The natural colour of the skin of these islanders is almost as white as that of Europeans; but when much exposed to the sun and air, it becomes brownish. This is particularly the case among the lower classes. Captain Wilson says, that many women who at his arrival were nearly as white as Europeans, by coming perpetually on board his ship in the burning heat of the sun became in time quite brown and sunburnt. The new-born children are nearly as white as in Europe. The women of the higher classes are not much less careful of their complexions than our belles, and in order to preserve them, live as much as possible under shelter from the scorching sun. Their garments are of a cloth made of the bark of the paper-mulberry tree; and when they go abroad they hold a green bough or a banana leaf over their heads by way of parasol. Hence many of them are not darker than a European brunette.

So strong is the desire to please among the women of these islands that they have even found out a method of bleaching their skin, when much tanned, in a few days. They generally have recourse to this expedient before any great festival or public sports that they may attract more notice and admiration. To this end, the whole body is rubbed with the juice extracted from the leaves of three different plants, which at first turns the skin quite black, so that for five or six days the persons undergoing the operation cannot leave the house: they then wash themselves well with fresh water, which takes off the black colour and leaves the skin of its natural whiteness.

Another custom, which equally prevails among both sexes, is to rub themselves all over with a sort of ointment, made of cocoa-nut oil mixed with the juice of various plants particularly a species of hibiscus, to give it a yellow colour. This shining yellow dye, in the opinion of these islanders, heightens personal beauty, particularly in men.

It renders the skin very soft and smooth and prevents strong perspiration, which in this hot climate very much weakens the constitution.

The dress of both sexes consists of a piece of cloth fastened round the waist, in the manner shewn in the opposite engraving of one of the natives of the Magdalena with a club. The women have also a large piece of cloth thrown over them, but less from modesty than to prevent the sun from injuring their complexions.

When the natives of Nukahiwa swam off to Captain Krusenstern's ship, the women, according to Langsdorff's account, came on board quite naked; for though none of them left the land with-

Native of La Magdalena. (*Marquesas Islands.*)
with a Club.

Pub. at R. Ackermanns. London, 1824.

out so much clothing as a large green leaf, yet this light garment was generally lost in swimming any distance. By a few only were the leafy aprons preserved, and luckily for them the Russians had no sheep or goats on board; otherwise those animals might have been not less eager to feast upon them than the Europeans to regale themselves on the bananas, cocoa-nuts and bread-fruit which these females brought in their hands. We are told by Captain Wilson in his account of the Missionary Voyage to the South Sea in the ship Duff, that an adventure of this kind happened to some visitors who came to pay their respects to himself and his crew. The knavish goats, says he, were guilty of a very great offence with regard to the poor young girls; for they would not leave them even the little clothing they had: they flocked round them to get at the green leaves till most of them were left entirely in their native beauty.

The practice of tattowing the body is common to all the islanders of the South Seas: but by none has the art been brought to such perfection as by the natives of the Marquesas. The regular designs with which the bodies of the men of Nukahiwa are punctured from head to foot, supplies in some measure the want of clothing, which, in so warm a climate, would be insupportable. Many are here as solicitous to obtain distinction by the symmetry and regularity with which they are tattowed, as among us by elegance of dress: and though no real elevation of rank is indicated by the superior style of these ornaments, yet as none but persons of the highest class can afford to be at the expense attendant on any refinement in them, it becomes in fact a badge of distinction.

The operation of tattowing is performed by persons who make a profession of it. For this purpose the artist uses the wing-bone of the tropic-bird, which is jagged and pointed at the end in the manner of a comb, sometimes in the form of a crescent, at others in a straight line, and of larger or smaller size according to the figures intended to be made. This instrument is fixed into a bamboo handle about as thick as the finger, with which the operator, by means of another cane, strikes so gently and dexterously, as scarcely to pierce through the skin. The principal strokes of the figures to be tattowed are first sketched upon the body with the same colouring matter that is afterwards rubbed into the punctures. These being so made that the blood and lymph ooze through the orifice, a thick dye, made of ashes of the kernel of the *tootooee*, or burning-nut, mixed with water, is rubbed in. A slight smarting and inflammation ensue; the wound then heals, and when the scar comes off after some days, the bluish or blackish figure appears.

When the young native of Nukahiwa approaches the age of manhood, the operation of tattowing is begun; and this is one of the most important epochs of his life. The artist is sent for; he contracts for a certain number of hogs, which is commonly regulated by the circumstances of the employer, as is the quantity of decoration by the pay. The sons of chiefs are shut up for several weeks in a separate house which is tabooed, to prevent the approach of all but persons whose attendance is required; and here they remain during the whole time that the operation continues. All women, the mother not excepted, are forbidden to see the youth while the taboo remains in force. Both the operator and his patient are fed on the very best kinds of food. In the first year only the groundwork of the principal figures on the breast, arms, back and thighs is laid: and in doing this the first punctures must be completely healed and the eschars come off before new ones are made. Each mark takes three or four days to heal; and the first sitting, as it may be called, commonly lasts three or four weeks. The annexed plate represents a young man who has been partially tattowed.

While the patient is undergoing the

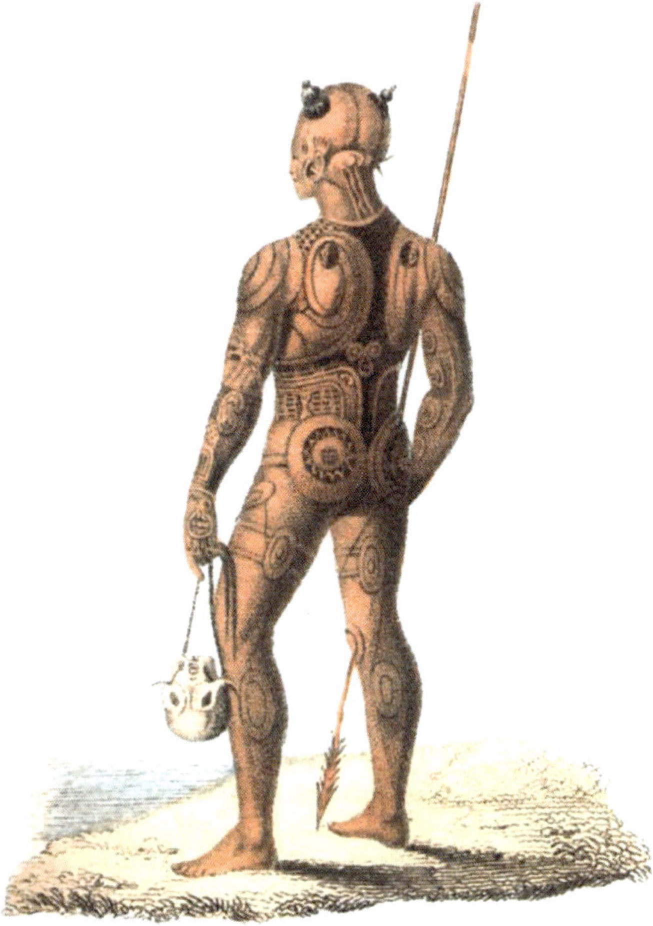

YOUNG MAN of the MARQUESAS,
not completely tatowed.

Pub. by R. Ackermann London, 1824.

operation, he must drink but little, for fear of inflammation, and he is not allowed to eat early in the morning, but only at noon and in the evening. When once the decorations are begun, some addition is constantly made to them at intervals of from three to six months; and it is not uncommon to proceed thus for thirty or forty years before the process of tattowing is completed. Some old men of the higher class are punctured over and over to such a degree, that the outlines of each separate figure are scarcely to be distinguished, and the body has an almost negro-like appearance. This, according to the notions of these people, is the height of perfection in personal ornament, probably because the cost has been great, and it therefore indicates a man of wealth and consequence.

The tattowing of persons of middling station is performed in houses erected for the purpose by the operators, and tabooed by authority. The poorer islanders, who live chiefly on bread-fruit, and have nothing else to give, are operated upon by novices in the art, who take them at a very low price as subjects for practice; but their performances are easily distinguished even by a stranger from those of an expert artist. Some of the lowest class, owing to poverty, are not tattowed at all.

The figures with which Cabri, a Frenchman, found at Nukahiwa by the

Vol. I. p. 223

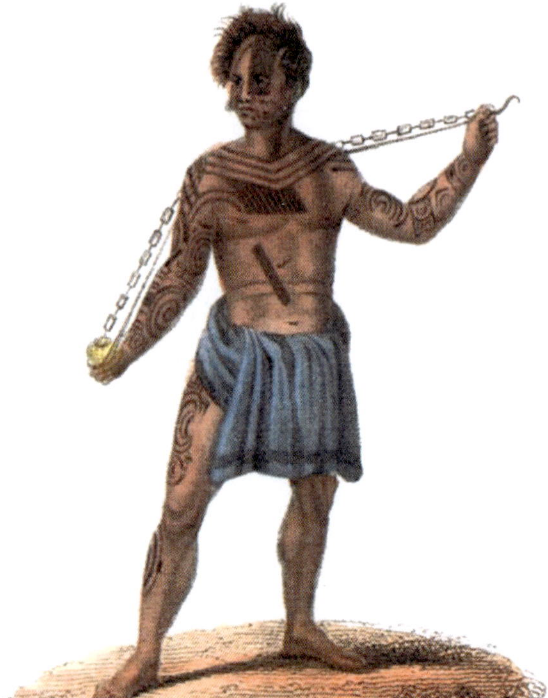

JEAN BAPTISTE CABRI.

Pub by R. Ackermann. London. 1824.

Russians was marked, seem to have been the work of some inferior artist. They are represented in the opposite plate. This man who had resided some years in the island, was brought to Europe by Captain Krusenstern, and returned to his native country, where he became an object of great curiosity.

The women of Nukahiwa are very little tattowed. The hands are punctured from the ends of the fingers to the wrist in a variety of patterns, which give them the appearance of wearing gloves. The feet also of many are so tattowed as to look like highly ornamented half-boots: long stripes are sometimes to be seen down the arms of the women and circles round them, which have much the appearance of the bracelets worn by European ladies. Some have likewise their ears and lips tattowed. The women are not like men shut up in a tabooed house while undergoing the operation: it is performed without ceremony in their own houses or wherever they please.

The figures with which the body is tattowed are chosen with great care, and appropriate ornaments are selected for the different parts. They consist partly of animals, partly of other objects, which have some reference to the manners and customs of the islands. On close examination curved lines, diamonds and other designs are often perceptible between rows of punctures which very closely resemble the ornaments denominated Grecian. Over the whole body the utmost symmetry is observed. The head is tattowed in every part: the breast is commonly ornamented with a figure resembling a shield. On the arms and thighs are stripes, sometimes broader, at others narrower, and in such directions that these people might be presumed to have studied anatomy and to be acquainted with the course and dimensions of the muscles. On the back is a large cross, beginning at the nape of the neck and ending with the spine. On the front of the thighs are often figures which seem intended to represent the human face; and on each side of the calf of the leg is an oval figure which produces a very good effect. In short, the whole displays much taste and discrimination, as may be seen from the annexed specimen of a distinguished native of Nukahiwa in this species of naked full-dres .

These people have a great variety of personal ornaments, but none that seems to be worn as a particular mark of distinction. Their principal head-dress, and what appears to be their chief ornament, is a sort of broad fillet curiously made of the fibres of the husk of the cocoa-nut: in the front is fixed a mother-of-pearl shell, wrought round to the size of a tea-saucer; before that another smaller of very fine

MAN OF DISTINCTION,
OF THE MARQUESAS.

Pub. by R. Ackermann London 1824.

tortoise-shell, perforated into curious figures; also before and in the centre of that is another round piece of mother of pearl about the size of half a crown, and before this another piece of perforated tortoise-shell as large as a shilling. Besides this decoration in front, some have it also on each side but in small pieces; and all have affixed to them the tail feathers of cocks, or of tropic birds, which, when the fillet is tied on, stand upright, so that the whole together makes a very sprightly ornament.

They frequently cut off all the hair except from two small circular spots just above the ears, and then tie the long hair growing on these places into bunches, which give them at a little distance the appearance of having horns. This fashion is represented in the engraving of one of these islanders with a sling. The same custom is almost universal among the children of the Chinese. Holes are bored in the ears large enough to admit a substance a third of an inch thick: a muscle shell, of an ounce weight, to which is attached the fang of a hog, or a light oval piece of bread-fruit wood is the favourite ornament for the ears. Large iron nails, a small stick about two inches long, and other trifles are also stuck into the ears.

Round the neck they wear a kind of ruff or necklace made of light wood,

Vol. I. p. 288.

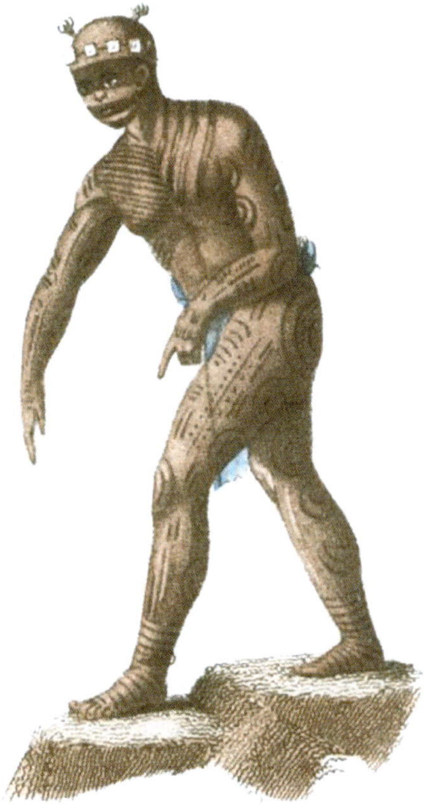

Native of La Magdalena. (*Marquesas Islands.*)
with a Sling.

Pub. by R. Ackermann, London. 1824.

the outer or upper side of which is covered with small red and black seeds of the *abrus precatorius,* which are fixed on with gum: they also wear bunches of human hair fastened to a string and tied round the legs and arms. All the above ornaments, however, are seldom seen on the same person.

Red feathers, or feathers naturally mottled, for they readily distinguish any colour given to them by art, are highly esteemed. To procure the tail feathers of the tropic-birds, they climb up the steep rocks at night and take the birds, while asleep, plucking the feathers without killing the birds, that they may repeat the same operation when the feathers are grown again.

The habitations of the people of Nukahiwa are commonly about twenty-five feet in length, and six or eight in breadth, with a partition across the middle. The wall in the rear is ten or twelve feet high; that in front not above three or four. They are formed of four strong posts, fixed in the ground at the corners, to which horizontal poles are attached. The sides are composed of bamboo canes of equal thickness, placed perpendicularly about half an inch asunder, and lined internally with leaves of the cocoa-palm and some species of fern dried. The roof is covered with several layers of leaves of the bread-fruit tree, which turn the heaviest rain. The entrance is in the lower wall in front. Langsdorff justly expresses his astonishment that not only here, but among all uncivilized nations, the entrance to their dwellings should be so disproportionately low. In cold climates, indeed, inhabited by a pigmy race, a good reason may be assigned for this; inasmuch as the smaller the aperture, the more easily the cold may be kept out: but it is incomprehensible how such a custom could have become universal among the tall and robust inhabitants of warm regions, to whom it must be a great inconvenience.

The best houses are raised on a platform of quadrangular smoothed stones, which sometimes extends several feet in front of the house; rendering the habitation more dry and giving to it a handsomer appearance. The skill with which they put together stones of such magnitude is astonishing: they are of such size as to require ten or twelve men to move them, and united without any kind of cement, so as to form absolute Roman walls. In erecting a new house neighbours mutually assist each other. The affluent frequently have huts in different parts of the valley where they reside, which may be taken down and removed in a few days.

Both sexes assist in the building of the larger dwellings, capable of accommodating a numerous family. But when a man, without the aid of his wife, collects the stones that are to serve for the ground-plot of his house, the building erected upon it is tabooed, that is to say, the women are prohibited from entering it. Every person of distinction has at least one such tabooed house, which is commonly at a little distance from the ordinary dwelling. He suits it entirely to his own convenience and has above all an eating-room, where apart from his wife he may eat swine's flesh undisturbed—for this is a food of which the women are rarely permitted to partake, and when they are, it is only by the special favour of the men.

Every new-built house must be consecrated by a priest, who holds on the occasion a discourse unintelligible to the people. He is then entertained with the best and sleeps the first night in the house, which is thus secured for ever against evil spirits.

The interior of the house is kept very clean: it is divided into two unequal parts. In the first, which is the smaller, the stone pavement only is to be seen; but the other is strewed with soft grass, over which are laid straw mats, and on these sleep, all the inhabitants of the house without distinction of age or sex The walls are hung round with domestic utensils, such as calebashes of different sizes, cocoa-nut shells, fishing-nets, lances, slings, stilts, battle-axes, hatchets, sundry ornaments, drums and various other articles.

The chief difference between the inhabitants of the Marquesas and those of the Society Islands seems to consist in their different degrees of cleanliness. The former do not, like the latter, bathe two or three times a day, or wash their hands and face before and after every meal; and they are moreover slovenly in the manner of preparing their food. Captain Cook saw the natives of Hood's Island make a batter of fruit and roots diluted with water in a vessel that was covered with dirt, and out of which hogs had just been eating, without washing it or their hands which were equally filthy.

Their mode of making cloth from the bark of the paper mulberry-tree differs but little from that practised at Otaheite. This occupation is left entirely to the women; while the principal business of the men consists in making their household utensils and weapons of war. These are wrought with great care and taste: their lances, spears and clubs are ornamented with carving, or with the hair of enemies ingeniously woven together. Their slings are made of the fibres of the cocoa-nut and other plants. About the morais are to be seen figures evidently designed to represent the human form, but which prove that the carvers have made no great progress in their art.

Native of La Magdalena. *(Marquesas Islands.)*
with a Fan.

Pub. by R. Ackermann London, 1824.

The fans for defence against the sun or wind, display considerable neatness and ingenuity: their form is shown in the annexed plate, and still more distinctly in that of a man completely tattowed: and the calebashes used for domestic purposes are decorated with human and other bones and net-work.

The people of distinction at Nukahiwa, as among some of the Asiatic nations, suffer the finger-nails to grow very long, to shew that they are not accustomed to labour. To give a refusal or a negative answer, they stretch out the tongue. Pointing with the fore-finger is a demonstration of friendship; and the mode of salutation when two friends meet, is to press the ends of the noses together.

The inhabitants of the Marquesas are canibals. It is an established fact that they eat their friends when pressed by famine, and their enemies from hatred or custom. The *tauas*, or priests, go still farther, and frequently regale themselves with human flesh merely for the delight they take in it. For this purpose they affect to be under the influence of a spirit, and after various grimaces and contortions fall apparently into a profound sleep; but they take good care that this shall always happen in the presence of numerous spectators. After pretending to sleep a short time they suddenly wake, and relate to the people around them what the spirit has during their trance directed to be done. They are commanded to seize and bring to the priest some person of whom a general description is given. They go accordingly and lie in wait near the place that is specified, and the first person coming that way and bearing any resemblance to the object described by the priest is secured, conducted to his morai, and dispatched for the purpose of being devoured.

In case of the illness of a *taua*, one, two or three inhabitants of a valley with which they are at enmity, according to the degree of his illness, are taken prisoners to be sacrificed to his luxurious appetite. Should the *taua* recover, it is well; should he not, the sacrifice is repeated; and if he nevertheless dies the whole valley participates in the event. The inhabitants seize and secure as many prisoners as they can, not by secret ambush but by open violence, considering them as enemies merely perhaps because they dwell on the other side of the river.

The man who kills an enemy has the head for his portion. He cuts it off immediately, parts the skull asunder at the sutures, and swallows the blood and brains on the spot. The flesh being carefully removed, the skull is then ornamented with hogs' bristles, and the lower jaw ingeniously fastened to it with thread made of the fibres of the cocoa-nut. It afterwards serves as a token of valour, for which purpose it is suspended from the cloth worn round the waist.

In open warfare, when they go to battle, they have their hands and feet ornamented with feathers, and their heads covered with a sort of veil woven out of the fibres of the cocoa-nut, which a stranger would take for an article of finery rather than part of their military accoutrements. The most distinguished heroes then have the skulls of enemies formerly slain bound to the hip or the foot. Their arms are slings, lances, javelins and clubs of casuarina wood. During the fight they leap and jump about, making all sorts of movements, and frequently quit the field when they have killed only one of their adversaries.

When they lurk singly after an enemy, it is commonly in bad weather, during heavy rain, as a greater number of persons are then employed in fishing, and the victim cannot so easily hear and see the foe in his ambush.

Young unmarried females are not subject to any restriction in their intercourse with the other sex. In a wife, however, infidelity is chastised with stripes, or the husband may discard her altogether, while the seducer is privately punished at the pleasure of the latter. As soon as a man and woman have mutually agreed to live together, presents are exchanged between the bridegroom and the parents of the bride. The union lasts as long as harmony prevails between the parties; but if either wishes to dissolve the contract, they are free.

At the marriage of the daughter of a person of distinction a certain number of hogs are killed, and all the friends and acquaintance are invited to the festivities which last two or three days. In case of infidelity her husband has a right to transfer her, even against her will, to any man who will take her. The rich may have as many wives as they please or can maintain; but it is more usual to have only one.

When a person dies, the corpse is

immediately washed clean, laid upon a sort of bier made of bamboo canes, and covered with several pieces of new cloth. Half the hogs belonging to the deceased are then killed and roasted: the *taua* and other friends are informed of the event, and invited to the funeral. Before the arrival of the guests, a large piece of cloth is spread out behind the bier on which the corpse is laid, and several stakes are stuck up with pieces of fine white cloth upon them to mark the taboo-place. The *taua*, who brings with him several large drums, performs the funeral ceremonies. These consist in a long harangue delivered by him in a language wholly unintelligible, the drums beating all the time. The guests then fall to feasting and keep up the banquet till every morsel of the hogs, bread-fruit, and other provisions, is consumed. The *taua* has the heads of the animals for his portion, and a hind-quarter must be reserved for the chief, who is invited to all funerals in his valley, and should he not attend in person it must be sent to his house.

During this time the nearest relations watch day and night by the corpse, and rub it well every day with cocoa-nut oil, which preserves it from corruption, and as the process sometimes lasts for weeks, changes it to an absolute mummy. The body is at length wrapped in cloth which has been steeped in the same oil, and placed upon a bier in the morai or family cemetery, which women are not permitted to enter.

Several kinds of magic or enchantment are practised here not only by the *tauas*, but by other individuals. One of the most remarkable is the art of bringing illness upon such as have offended them. For this purpose a preparation called *kacha* is requisite. The person who would make a *kacha* must wash his whole body very clean, abstain for a certain time from food, and live in a tabooed place. The *kacha* itself is a small pouch or bag made of stuff woven with cocoa-nut fibres and other threads, in which are tied up the skin of a fresh-killed lizard, various sorts of plants, a little stone of a particular form, a small piece of bamboo cane and several other articles. But the principal point is to seek out some one of the enemies of the man against whom the charm is directed, and to procure from him some of his hair, the remains of something he has been eating, earth on which he has spit, and a small piece of the cloth which he ties round his waist. Without these things the *kacha* would be inefficient.

Three of these enchanted bags are usually prepared: when ready they are well perfumed and buried separately in the ground in some retired spot. Should the person whom they are designed to injure chance to fall ill, he, under the impression that his illness is the effect of enchantment, seeks, with the aid of a *taua*, to discover the author of the mischief. The suspected person is interrogated, and if he be really the composer of the charm, he must acknowledge the fact, otherwise he is guilty of transgressing a taboo, and immediate death would, according to his belief, ensue. The sick man, on discovering his foe, presents him with one or more hogs, to induce him to break the spell, which is done by digging up the *kachus*. If the first present prove insufficient, so that the patient does not recover, he must renew his bribe; if the second will not do, he must try a third, after which the last *kacha* is taken out of the ground.

If the enchanter be sufficiently paid, the patient recovers; if not there is no hope for him; as the spirit then sides with his enemy and cannot be appeased.

The fear of enchantment and of the influence of spirits is the principal curb by which the passions of these people are kept under any kind of restraint. Roberts, an Englishman, who had been some years at Nukahiwa at the time of Krusenstern's visit, assured the Russians, that he often threatened the natives with setting his spirit to work, and that by this menace he could generally make them do whatever he wished, because they were afraid that he would bring upon them a number of European ships to destroy them at once.

According to the popular belief, the *tauas* are enabled by their knowledge of magic to discover the causes of many diseases. When any one finds himself indisposed, he sends for a *taua*, who passes his hand three times over the patient's breast and pretends by this method to drive away the spirit. So completely had the belief in this superstition possessed itself of the mind of Cabri, the Frenchman already mentioned, that he assured Mr. Langsdörff, he had himself heard the spirit in the palm of the *taua's* hand, where he sometimes whistles very loud. He added, that when the *taua* enquires why he has made the man ill, the spirit answers in a language known to the priest only; but according to his account, the reason generally assigned is, that the patient has transgressed a taboo, perhaps taken some bread-fruit or cocoa-nuts from a tabooed tree. He is then required to specify the number of hogs which must be given by the offender to atone for his transgression, and to restore the taboo, without which he cannot recover. The number of hogs demanded is commonly in proportion to the wealth of the offender. These peace-offerings are carried to the *taua's* morai and there consumed.

If any thing be stolen, for instance,

a hog, and the person robbed has a suspicion who is the thief, he in revenge taboos the hogs or other property of the latter. This is done by giving names to the animals or trees; by which means, according to the ideas of these people, they are bewitched or enchanted: and thus the suspected person is sometimes compelled to leave the place and all his possessions and settle elsewhere. The hogs that have been thus charmed must never be killed. Persons of either sex may charm the finest of their trees, which then become tabooed, and their property is thus rendered secure.

Every child from the moment of its birth inherits from its parents at least one bread-fruit tree, and this tree is tabooed even to the father and mother. If the parents are so poor that they have not a tree to settle upon the child, one is planted for it immediately; and thus a provision is made for its future maintenance; since one or two of these trees are sufficient to support an individual the whole year round.

In times of plenty these gay people have a variety of amusements. At the season when the bread-fruit is ripe, the chiefs give entertainments, feasting the people for some time on hogs, cocoa-nuts, bananas and many species of roots. The principal of these assemblies are the dancing festivals. In their dances the performers make many springs and pantomimic gestures, with quick movements of the hands and arms but without moving much from one spot.

At these dances the company appear in their most costly ornaments, the principal of which are composed of the tail-feathers of the tropic-bird. Six rings decorated with them are put on each middle finger of the dancers, and with some of these feathers, mixed with other kinds and hair, various ornaments are ingeniously made for the hands, feet, hips, head, neck and ears.

The place where the dance is held is the most level part of the valley; it is seldom less than two hundred yards long, paved with flag stones, laid as even as if done by a European workman. The inhabitants of the neighbouring valleys are invited to participate in these diversions; but they never come unarmed, and keep all together on one side of the dancing ground, that in case of emergency they may be ready to stand on the defensive. They commonly carry with them a sling, but it is bound round the head so as to have the appearance of an ornament rather than of a weapon. These guests in fact are not always safe from attack, since festivities of this sort have been employed as a pretext to draw victims into a snare and to seize them with the greater facility. If a *taua* takes it into his head to be influenced by the spirit or to fall into a trance, this is sufficient to put an end to all conviviality and to occasion hostilities.

The music at these festivals consists of a wild sort of cry, and the beating of several drums, some of them of large dimensions. They are in the shape of an upright cylinder or cask, and are from four feet to four feet and a half high, and eighteen inches or two feet in diameter: the skin of a shark is stretched over the top, and the workmanship is extremely neat. The cylinder of the drum is made of the hollow trunk of a tree; and at the lower end, which is smaller than the other, it has holes like the sounding-board of a stringed instrument.

Drums are used principally at funerals, but never with the dancing. These and the war-trumpet, which, with the cries of the warriors, urge the people on to battle, are the only musical instruments of the islanders. The trumpet consists of a large shell, the Triton's horn, which is adorned with the hair of a slaughtered enemy, and at the small end of which is fixed the shell of an oil-nut as a mouth-piece.

Next to dancing a favourite and we may say peculiar amusement of these people is running on stilts: and perhaps no nation shows so much dexterity in this exercise as the natives of the Marquesas. At their public festivals they run in this way for wagers, each trying to cross and throw down his antagonist. They run on the dancing-place, the smooth pavement of which renders the task the more difficult. Children of eight or ten years old are perfect adepts in this exercise, the principal object of which seems to be amusement.

The dexterity of these people in swimming also excites astonishment. They appear in fact to be a sort of amphibious animals, capable of living just as well in the water as on land. They will throw themselves from high steep rocks into the sea; remain for a considerable time in the same place as if they were standing upright, with the head and shoulders above water; and bring a number of things for barter, which they hold above the surface to keep them from getting wet. Some of them, says Langsdorff, would swim about the ship the greatest part of the day without appearing tired. Mufau who has been already mentioned, though he had never before been on board a European ship, would of his own accord run up the main-mast several times successively and throw himself into the sea, to the great astonishment of the spectators. It was impossible to see without horror and amaze, how he would spring from such a height and balance himself in the air for some seconds with his legs drawn up and his head erect. Such was the force of the fall and the weight of his body, that several seconds elapsed before he again rose to the surface.

END OF VOL. I.

J. Green, Leicester Street, Leicester Square.

Printed in Dunstable, United Kingdom

91222218R00203